SPAD FIGHTERS
in action

By John F. Connors
Color by Don Greer
Illustrated by Perry Manley

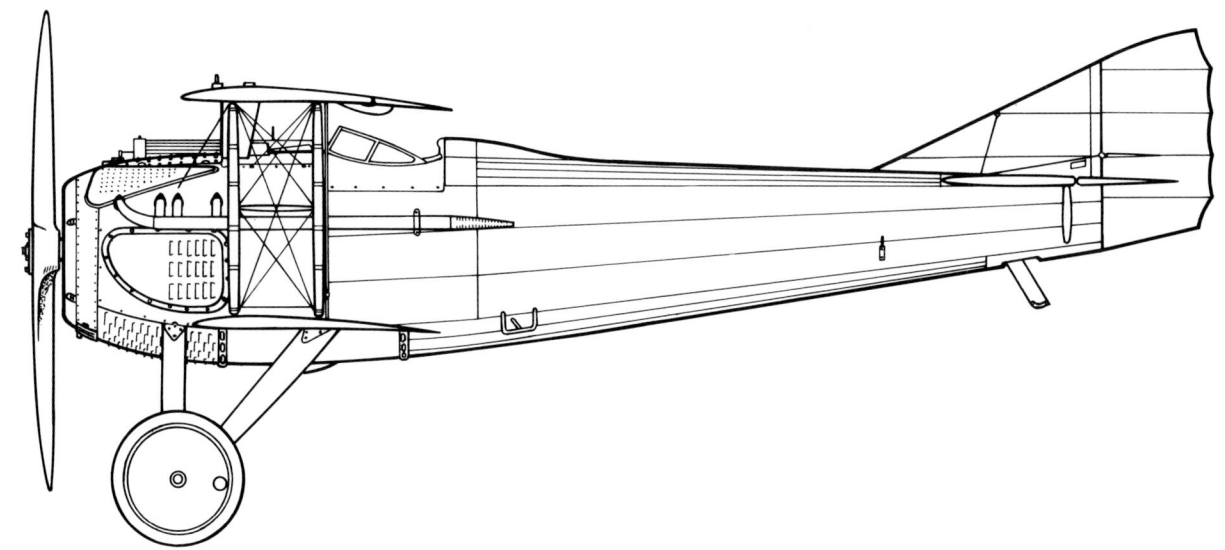

Aircraft Number 93
squadron/signal publications

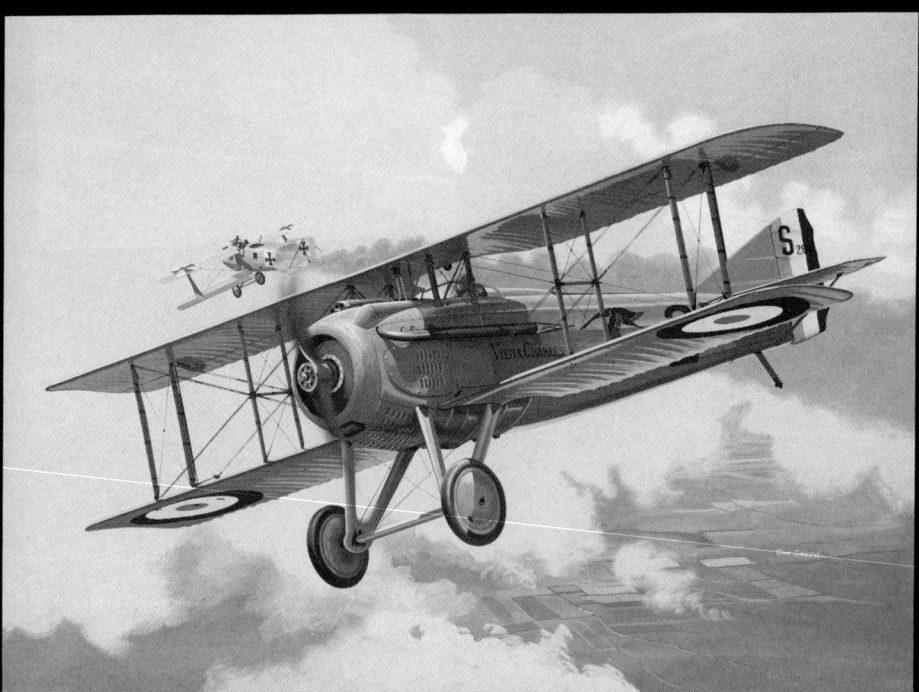

CAPT Georges Guynemer of Spa.3 engages a German Roland C.II observation aircraft over the front during early 1917. Guynemer's Spad 7 carried the name *Vieux Charles* on the fuselage side in Black.

COPYRIGHT © 1989 SQUADRON/SIGNAL PUBLICATIONS, INC.
1115 CROWLEY DRIVE   CARROLLTON, TEXAS 75011-5010
All rights reserved. No part of this publication may be reproduced, stored in a retrieval system or transmitted in any form by any means electrical, mechanical or otherwise, without written permission of the publisher.

ISBN 0-89747-217-9

If you have any photographs of the aircraft, armor, soldiers or ships of any nation, particularly wartime snapshots, why not share them with us and help make Squadron/Signal's books all the more interesting and complete in the future. Any photograph sent to us will be copied and the original returned. The donor will be fully credited for any photos used. Please send them to:

Squadron/Signal Publications, Inc.
1115 Crowley Drive.
Carrollton, TX 75011-5010.

Acknowledgements

I am indebted to several institutions and individuals which helped make this book an accurate history of the Spad World War One fighter aircraft. My thanks first of all to J. M. Bruce for making available his own Spad information, for direction as to where to locate additional information, and for being a constant source of encouragement and advice. I am grateful to Robert Sheldon for the use of his personal library on World War I aviation. I also appreciate the assistance of Ellic Somer. His *Spad Newsletter* keeps everyone interested in the aircraft abreast of the latest information. My thanks also to Russell Hansen, who was very generous with his Spad material.

Photo Credits

My thanks to the following for the use of photographs, documents, and other materials for this book:

Musee de l'Air
Royal Air Force Museum
Roberto Gentilli
Robert Sheldon
Russell Hansen
Public Record Office
Section Historique
de l'Armee de l'Air (SHAA)

Cross and Cockade Journal
Imperial War Museum (IWM)
National Air and Space Museum (NASM)
J.M. Bruce/G.S. Leslie Collection
Dennis Hylands
Ernest R. McDowell

Dedication

To Fred and Betty Jo

New production Spad 13s and Spad 7s await delivery to fighter units at a French aircraft depot during the Spring of 1918. It is believed that most of the aircraft here were built by Bleirot, one of the sub-contractors in the Spad 13 program. (Musee de l' Air)

Introduction

One of the most illustrious names in French aviation during the years just prior to the First World War was that of Armand Deperdussin, a wealthy French industrialist who had realized that aviation had great potential for business. During 1910 he formed the *Societe Provisoire des Aeroplanes Deperdussin*. According to customary French business practice, the aircraft produced by this firm could have been referred to as SPADS, an acronym composed of the initials of the company. During the company's early years, however, aircraft built by the firm were called simply, Deperdussins. These included the Deperdussin monoplane racers which exhibited very advanced streamlining for their day and won the Gordon Bennet Trophy for France during 1913. During that race a Deperdussin, flown by Maurice Prevost, attained a speed of 125 mph.

Unfortunately, during 1914, Armand Deperdussin became involved in a number of questionable financial dealings which became so serious that the financial stability of France itself was threatened. On 5 August 1914, he was arrested on a charge of embezzlement, however, final judgment in his case was not handed down until 30 March 1917, at which time he was convicted and sentenced to five years imprisonment. He was immediately released under France's First Offenders Act, but never again became involved in aviation.

Deperdussin's legal problems caused great concern for those involved with his aircraft company which was completely legitimate and at the peak of its success. The board of directors decided that the best course of action was to name a new director. They chose Louis Bleriot, a French national hero who, on 25 July 1909, had made the first flight across the English Channel. More importantly — Bleriot's reputation was impeccable.

A group of French industrialists, headed by Bleriot, acquired the assets of the Deperdussin works and set up a new company in August of 1914. The name of the company was changed to *Societe Anonyme pour l'Aviaton et ses Derives* (Society for Aviation and its Derivitives), in a seemingly deliberate effort to keep the initials S.P.A.D. (although the reason for this is unknown). From that point forward all the company's products were called by the acronym Spad.

Bleriot wisely decided to keep as much of Deperdussin's talented staff as possible, especially Louis Bechereau, the chief engineer responsible for the Deperdussin racers. During the first few months of the First World War, the S.P.A.D. works produced no aircraft for the French *Aviation Militaire*, although Bechereau had drawn up a number of designs.

The Spad A.2

The first military aircraft put into production by the S.P.A.D. works was the Spad A.2. At the time of its inception, the Allies had not yet perfected the machine gun synchronization system, and the A.2 represented one method of arming an aircraft with a forward firing machine gun.

The Spad A.2 was a tractor biplane featuring a long tapered fuselage and a streamlined nacelle mounted in front of the engine and propeller. This nacelle housed the observer/gunner who was equipped with a flexible Lewis machine gun on a tubular mounting. To provide cooling air for the engine, a set of screened air intakes were mounted on either side of the nacelle. To start the engine, the nacelle was detached from its upper mountings and swung down clear of the propeller. Once the engine was started, the nacelle was repositioned and locked in place. To protect the observer from accidentally coming in contact with the propeller, a wire mesh screen guard was mounted directly behind the cockpit. The pilot's cockpit was located in the main fuselage directly behind the wings.

The first prototype, designated the Spad A.1, first flew in May of 1915, and was

'MA JEANNE' was one of forty-two Spad A.2s that served with the French *Aviation Militaire*. The forward nacelle was movable and could be rotated downward to provide access to the engine. (SHAA)

This Spad A.2 served with the Imperial Russian Air Service and was fitted with skis for operations from snow covered airfields. This aircraft carries a rear view mirror attached to the rear fuselage cabane strut. (SHAA)

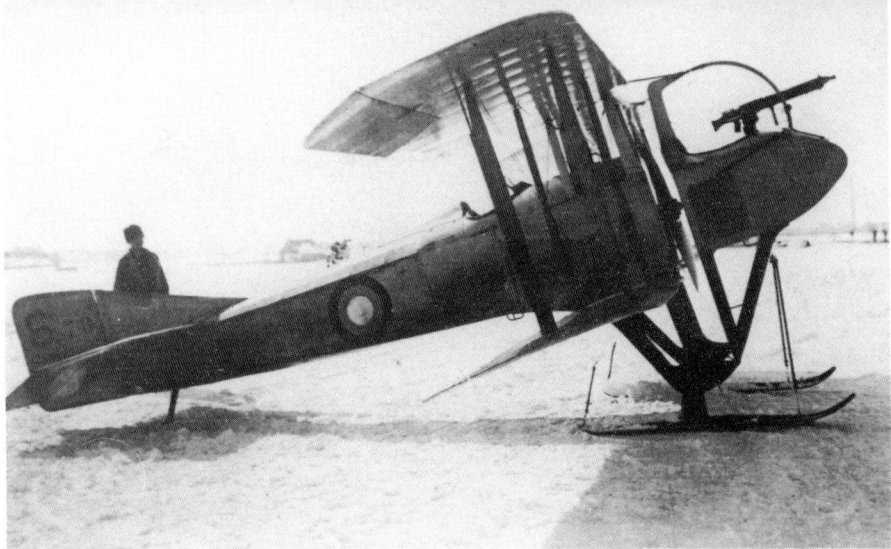

The Russians flew the Spad A.2 and its successor the Spad A.4 long after the French had retired them — chiefly because there was a chronic shortage of aircraft within the Russian Air Service. This aircraft (serial S.79) is armed with a 303 Lewis machine gun. (RAF Museum)

powered by a nine cylinder 80 hp Le Rhone 9C rotary engine. The aircraft reportedly reached a top speed of 152 kph (95 mph). The A.1 did not enter production, as it was superseded by the improved Spad A.2. A British report dated 11 October 1916 gave the top speed of the Spad A.2 as 100 mph at 6,600 feet, with a time to climb of seven and a quarter minutes. Speed at 10,000 feet was reported as 96 mph.

A total of ninety-nine A.2s were built, forty-two being delivered to the French *Aviation Militaire*, and fifty-seven being exported to Russia for service with the Imperial Russian Air Service. A number of Russian Spad A.2s had their wheeled landing gear replaced with a ski landing gear for winter operations from snow covered fields.

The Spad A.2 was not a popular aircraft with its crews. The effectiveness of the propeller was limited by the observer's forward nacelle, and visibility from the cockpit was very restricted, particularly in the landing configuration. The A.2 had a brief career with the French *Aviation Militaire*, although the Russians (who were always short of aircraft) kept the A.2 in front line service for a longer period.

Although not a great success, the Spad A.2 had several design features which would be used on later Spad fighters. The wings were long and of extremely thin cross-section, braced as a single bay on each side. Because of their length and the narrow gap between them, the wing flying and landing wires formed acute angles. To keep the flying wires from vibrating excessively in flight, an additional set of interplane struts were fitted at the point where the flying wires crossed, giving the aircraft the appearance of a two-bay biplane. The ailerons were attached to the upper wings only and were operated by control rods which ran from the base of the control column, through the lower wings to external bell cranks at the base of the rear outboard interplane struts. These bell cranks were connected to vertical rods linked to the actuating levers of the ailerons.

Developments of the Spad A.2 included the A.3, a dual-controlled version with a gun fitted to both cockpits. Exact details of this aircraft are unknown and apparently only one was built. Eleven examples of the Spad A.4, first flown in February of 1916, were completed, ten being delivered to the Russians. The Spad A.4 was basically an A.2 airframe fitted with a nine cylinder 110 hp Le Rhone 9J rotary engine. The final variant was the A.5, built for the Military Competition of 1916. The A.5 was basically similar to the A.4, but was powered by the Renault 8 Fg engine.

The front nacelle of the Spad A.2 pivoted around its lower attachment points to the undercarriage. The upper support struts detached allowing the nacelle to be lowered for engine maintenance. The small wire screen at the rear of the cockpit was designed to protect the observer from the propeller. (SHAA)

Development

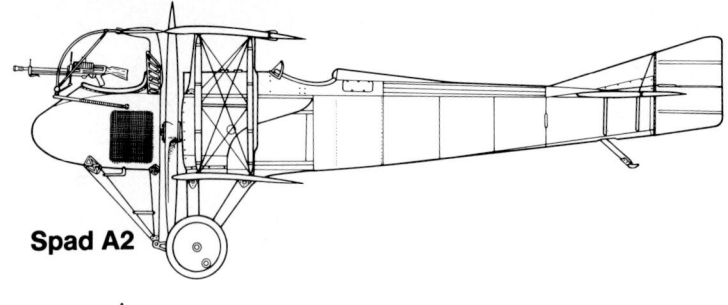

Spad A2

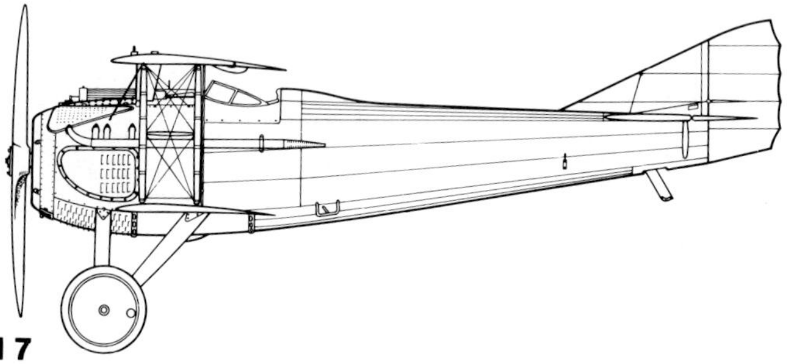

Spad 7

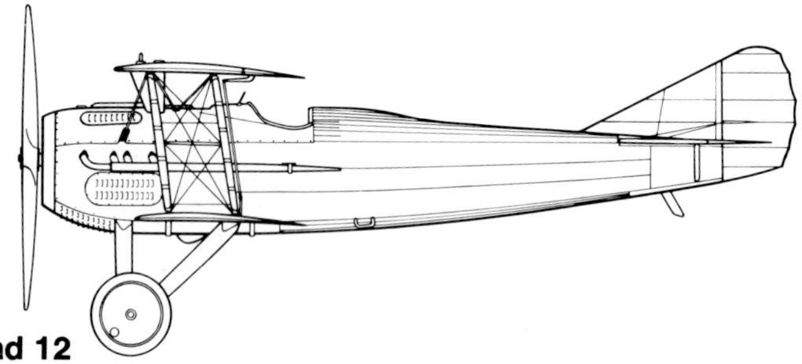

Spad 12

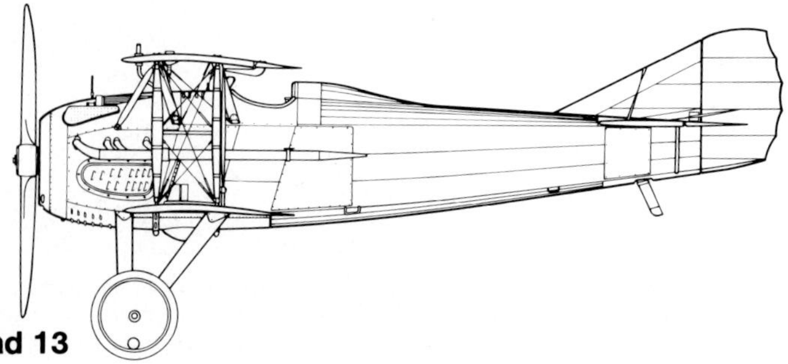

Spad 13

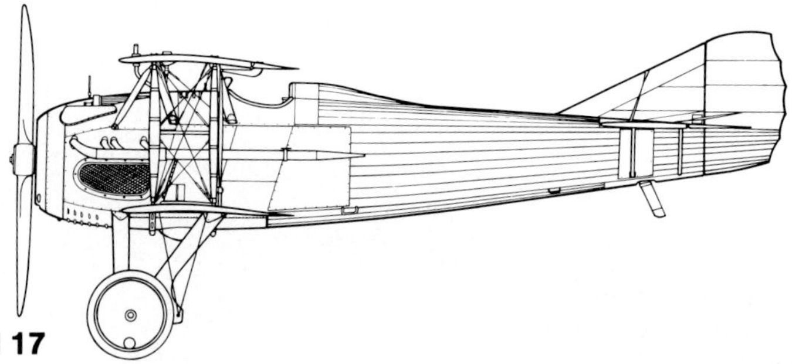

Spad 17

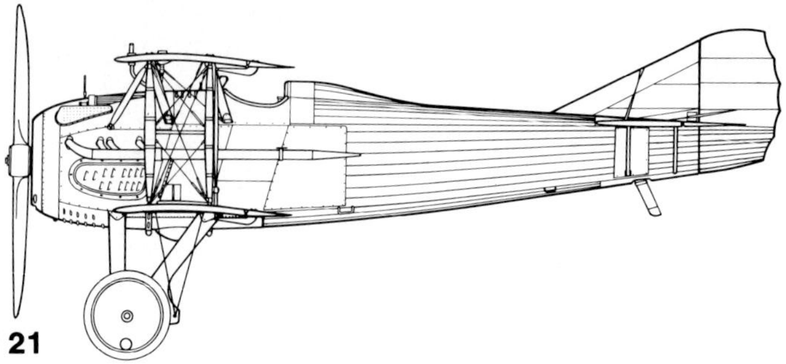

Spad 21

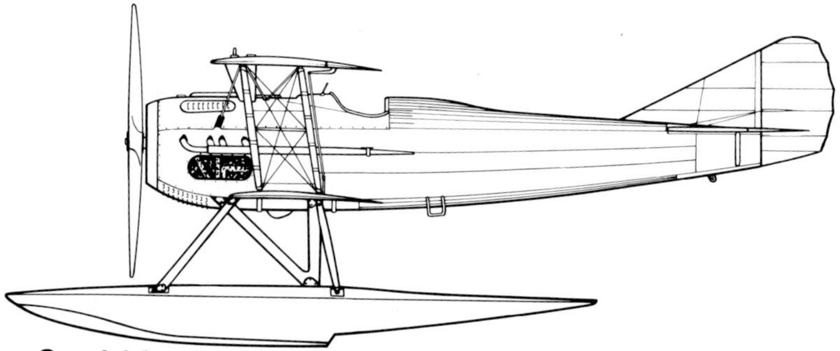

Spad 14

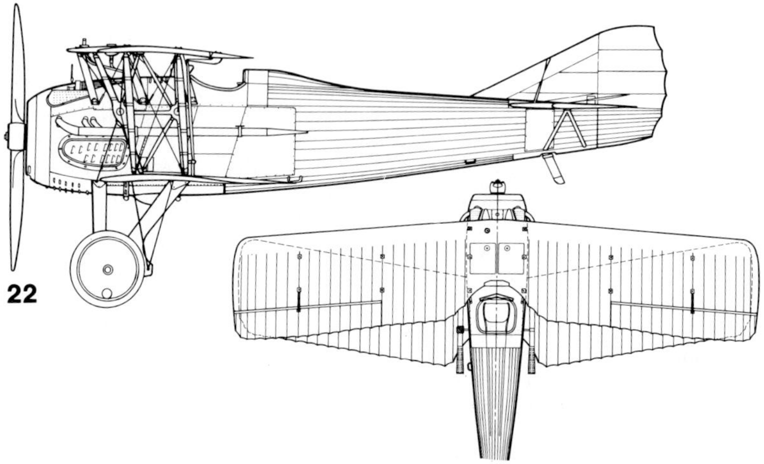

Spad 22

Spad 7.C.1.

One of the best power plants available during the early 1900s was the Swiss designed, Spanish built Hispano-Suiza V-8. This engine had gained an excellent reputation for power and reliability in high performance automobiles, and, during 1914, the engine's Swiss designer, Marc Birkigt, began work to modify the engine for aircraft use. The prototype engine was ready by February of 1915. It weighed 330 pounds and delivered 140 hp at 1,400 rpm. During July of 1915, a second prototype engine was tested at Bois-Colombes, where it delivered 150 hp at 1,550 rpm. French *Aviation Militaire* officials were impressed and ordered the engine into immediate production.

To meet a French General Headquarters' request for an aircraft designed around the 150 hp Hispano-Suiza, Bechereau designed a small, compact, single-bay, single-seat biplane under the company designation Spad V. The Spad V had a family resemblance to the earlier Spad A.2, especially in the profile of the rear fuselage and vertical fin, as well as in the thin cross-section wings. The Spad V retained the unique system of interplane bracing and aileron control which had proven successful on the Spad A.2. Armament was a single Vickers machine gun mounted in front of the cockpit, slightly off-set to starboard, synchronized to fire through the propeller arc by a Birkigt synchronizing gear.

The Spad V followed the standard construction techniques of the period, being made primarily of a wooden frame with fabric covering. The fuselage forward of the cockpit was covered with sheet steel panels, while the wing struts were of duralumin tubes faired into a streamlined shape with spruce moldings. The engine radiator was closely cowled. Fuel was carried in an under fuselage tank which was molded to fit the shape of the lower front fuselage. The fuel was transferred to a service tank in the center upper wing section by an engine-driven fuel pump. The close set landing gear was fixed to the fuselage with one piece legs made of carved laminated poplar sheets with a fixed axle beam running between them. The tail skid was of also of wood, tipped with a steel shoe.

Test flights of the prototype Spad V began in April of 1916, with company test pilot Bequet at the controls. The prototype attained a top speed of 196 kph (122 mph), with a time to climb to 3000 meters (9,840 feet) of fifteen minutes. Later that same month, and continuing into May, official French *Aviation Militaire* trials were carried out at Villacoublay. A Royal Naval Air Service report dated 11 November 1916, recorded a speed of 186 kph (116 mph) at 3,000 meters (9,840 feet), and 199 kph (124 mph) at ground level. Climb to 3,050 meters (10,000 feet) took some eleven minutes. The report also stated that one of the aircraft's chief merits was its diving ability, and further stated that one French pilot reported diving the machine "perfectly, comfortably and controllably at 400 kph (249 mph)."

The performance of the new Spad caused considerable excitement in the *Aviation Militaire* and an initial production contract was awarded on 10 May 1916 for 268 machines under the designation Spad 7.C.1.

Unfortunately, because of problems with the 150 hp Hispano-Suiza engine and the machine gun synchronization system, production was slow to begin. Further tests had revealed that a number of modifications were required, although the report, dated 11 November 1916, containing the recommendations of Capitaine le Reverend of the *Ministere de la Guerre* (War Ministry) to the *Commission des Marches (Chambre des Deputes)* does not specify what these modifications were.

By August of 1916 a small number of early production Spad 7s had been delivered to French *Escadrilles de Chasse* (fighter squadrons). For the rest of 1916, S.P.A.D. worked to iron out the problems with the Spad 7, and by the beginning of 1917, production aircraft finally began to be delivered to fighter units in larger numbers. Word about the new Spad

This aircraft is believed to be the Spad V, prototype for the production Spad 7. The aircraft lacks national markings, has short individual exhaust stubs, and a smaller windscreen than production Spad 7s. (IWM)

Spad Wing Bracing

Conventional Single Bay Biplane

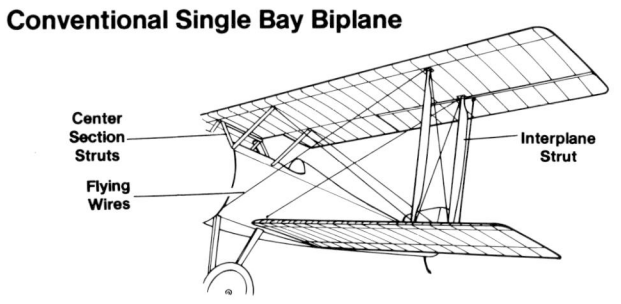

Spad 7

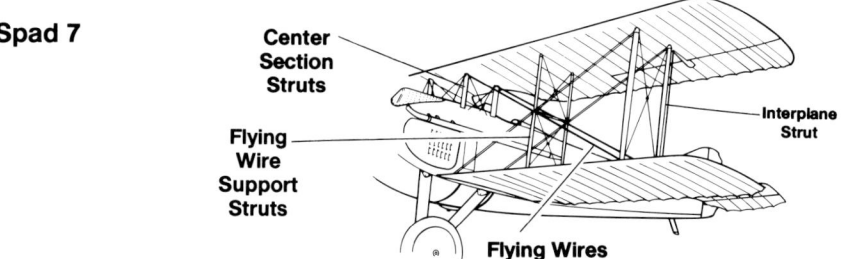

7

fighter had already reached the front line units and French aviators were anxiously awaiting the arrival of the new fighter. The Nieuport 17s, which equipped most front line units, were by then being outclassed by the current generation of German fighters and most pilots saw in the Spad a way of regaining equality, if not supremacy, in the air.

Officers of the *Aviation Militaire* were also anxious to see how the Spad 7 performed in combat against enemy aircraft. LT Armand Pinsard, who had recently been posted to *Escadrille* N 26 after escaping from a German prison, was asked by COL Joseph Bares to take one of the first production Spad 7s (possibly S.122) to the front and test it in combat. The exact date this aircraft arrived at the front is unknown, however, on 26 August 1916, Pinsard scored the first victory for the Spad 7.

Despite the delays in production and delivery, Spad 7s began to trickle to the front during late August, with some pilots collecting them personally from the testing station at Villacoublay. The first Spad 7 to be officially assigned to a squadron was aircraft serial S.112, flown by Second LT Paul Sauvage of *Escadrille* N 65. The second was serialed S.113 which was delivered to *Escadrille* N3. This aircraft became the first of three Spad 7s to be flown by the French ace S.LT Georges Guynemer. He flew the Spad in combat for the first time on 4 September 1916, and gained his fifteenth victory in it by shooting down an Aviatik. Guynemer felt an immediate love for the Spad and, in a letter dated 8 September of 1916, he reported to Bechereau that; "She loops wonderfully. Her spin is a bit lazy and irregular, but deliciously soft."

Most pilots considered the Spad 7 to be inferior to the Nieuport 17 as a dogfighter because it was heavier and lacked the maneuverability of the Nieuport. It was, however, much stronger, making it one of the best diving aircraft of the war. Additionally the Spad 7 could take a good deal more punishment that other fighters and still remain in the air.

One problem that continued to beset the Spad 7 was engine cooling. The firm manufacturing the engine radiators had run out of raw materials in October of 1916, causing further delays in production and maintenance problems with aircraft already in service. During the colder months, the engines could not be kept warm enough, and during warmer weather, they easily over heated. The latter problem was helped by enlarging the opening in the cowling slightly. As for the former difficulty, a variety of temporary fixes were tried in the field including several styles of radiator shutters. Eventually, a standardized arrangement consisting of nine vertical shutters with two horizontal mounting bars became standard.

The engine mounts of the Spad 7 were also reported to be weak. Engine vibration and rough landings often caused them to bend or break. This problem was solved by strengthening the mounts with a three-ply mounting plate and a steel brace on its weakest point.

In order to increase the output of Spad 7s, nine firms were subcontracted to assist S.P.A.D. with production. Still, the numbers of Spad 7s at the front were below expectations and delivery schedules were consistently not met. By February of 1917, six months after the Spad 7s introduction into service, 268 aircraft had been delivered. By 1 August, there were 495 in service, equipping more than fifty *Escadrilles de Chasse*. The Spad 7 was also gaining a measure of fame as the mount of such French aces as Fonck, Dorme, Madon, Nungesser, Deullin, Heurtaux, and Guynemer.

By the spring of 1917, the more powerful 180 hp Hispano-Suiza 8Ab engine became available for installation in the Spad 7. This engine provided the Spad 7 with a significant improvement in performance. Maximum speed was increased to 208 kph (129 mph), and climb to 5,000 meters (16,400 feet) now took twenty-two minutes. The first Spad 7 sent to the front with the new engine was S.254, which became the third Spad 7 flown by Guynemer, and the one in which he attained his most victories — nineteen in all. By April of 1917, the 180 hp Hispano-Suiza had become the standard engine in all production Spad 7s being built in French factories.

This Spad 7 (S.111) was the first aircraft off the production line. The aircraft was later modified with a wide radiator cowling opening. A collector pipe was fitted to the engine exhaust stubs. (Musee de l'Air)

S.P.A.D. Aileron Control System

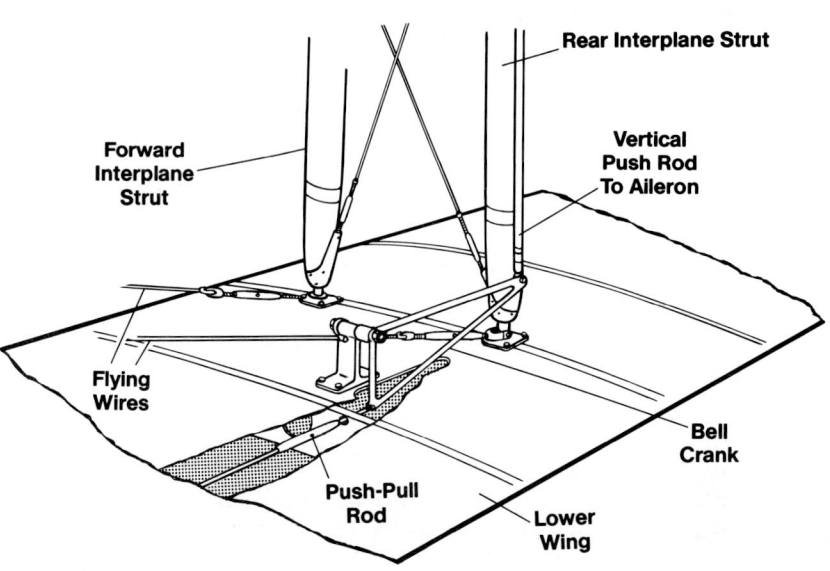

Other power plants were tried experimentally in the Spad 7. A 150 hp Renault was fitted to one aircraft. This engine was also a V-8, but with the cylinder banks at 60° (as opposed to 90° on the Hispano-Suiza). This installation required considerable modification to the aircraft's nose and cowling. Performance figures for this aircraft are unavailable; however, it is believed that it was not a success.

A supercharged variant of the Spad 7 was tested during 1917, fitted with a 190 hp supercharged Hispano-Suiza engine. Flight tests revealed that its performance was more or less equal to the standard Spad 7 with the 180 hp engine. One interesting note on the test flight report shows performance figures for a non-supercharged Spad with both a standard wing and a so-called "Bleriot" wing. Other reports speak of a 150 hp Spad 7 with a "flat" wing, presumably one with a revised airfoil section. This test reported identical performance to the "Bleriot" wing version and it is believed that the "flat" wing and "Bleriot" wing were actually the same. How the name Bleriot became associated with this wing is now unknown. Performance for the Spad 7 fitted with the Bleriot wing was slightly improved. Speed at 2,000 meters (6,560 feet) was 194 kph (121 mph), with a time to climb of five minutes, fifty seconds. Speed with the standard wing at the same height was 184 kph (114 mph), with a time to climb of six minutes and forty seconds.

The Spad 7 gave a good account of itself in combat, although its chief weakness was in the armament. The Spad 7 was armed with a single machine gun at a time when most German fighters were fitted with two. Spad 7s also served with distinction in the air services of a number of other Allied nations. It was both used and produced by the British. Spad 7s were supplied to the Belgians and formed the equipment of the *5e Escadrille* (later renumbered the *10e Escadrille*) of the *Aviation Militaire Belge*. Belgian ace Edmond Thieffry was the first Belgian pilot to receive a Spad and scored his first Spad victory on 12 May 1917.

Italy began to receive Spad 7s in March of 1917, with a number of aircraft being supplied to nine *Squadriglie* of the *Aeronautica del Regio Esercito*. Although most Italian pilots preferred the slower, and more maneuverable Nieuport types and the Hanriot H.D.1, many Italian aces were successful with the Spad. These included the leading Italian ace, Maggiore Francesco Baracca. Baracca was not bothered by the light armament of the Spad 7, "It doesn't matter if the 7 is equipped with a single gun," he said. "Provided you are a good fighter, a single gun is enough."

The American Expeditionary Force eventually obtained 189 Spad 7s, with deliveries beginning in December of 1917. Many of these served until the end of the war. In fact, two American units, the 138th and 638th Squadrons, did not receive the Spad 7 until October of 1918.

During the Spring of 1917, forty-three Spad 7s were sent to Russia, where they formed part of the equipment of No 1 Fighter Group, led by Alexander A. Kazakov. Additionally, approximately 100 Spad 7s were built by the *Aktsionyernoye Obschestovo Duks* company of Moscow. The fate of these aircraft after the Russian Revolution is unknown, except that one may have gone to Finland.

After the war, Spad 7s were widely exported and served in the air services of France (where it was the standard pilot's certificate test aircraft until 1928), Czechoslovakia, Finland (one), Greece, Poland, Portugal, Romania, Yugoslavia, Brazil, Peru, Siam, and the United States.

An early production Spad 7 (serial S.1837) on the airfield at Bar-le-Duc. The standard color scheme for early Spad 7s was overall clear dope on the fabric surfaces with the metal cowl panels painted Light Yellow to roughly match the doped fabric. The Yellow painted area extended to just behind the cockpit. (Musee de l'Air)

Ground crews bore-sight the Vickers machine gun of a Spad 7 into a hillside firing pit at a French airfield on the Marne during April of 1917. (Musee de l'Air)

Specifications

Spad 7.C.1

Wingspan	25 feet 8 inches
Length	19 feet 11¼ inches
Height	7 feet 2½ inches
Empty Weight	1,102 pounds
Maximum Weight	1,554 pounds
Powerplant	One 180 hp Hispano-Suiza 8Ab engine
Armament	One 303 Vickers machine gun
Performance	
Maximum Speed	129 mph
Service ceiling	21,490 feet
Range	217 miles
Crew	One

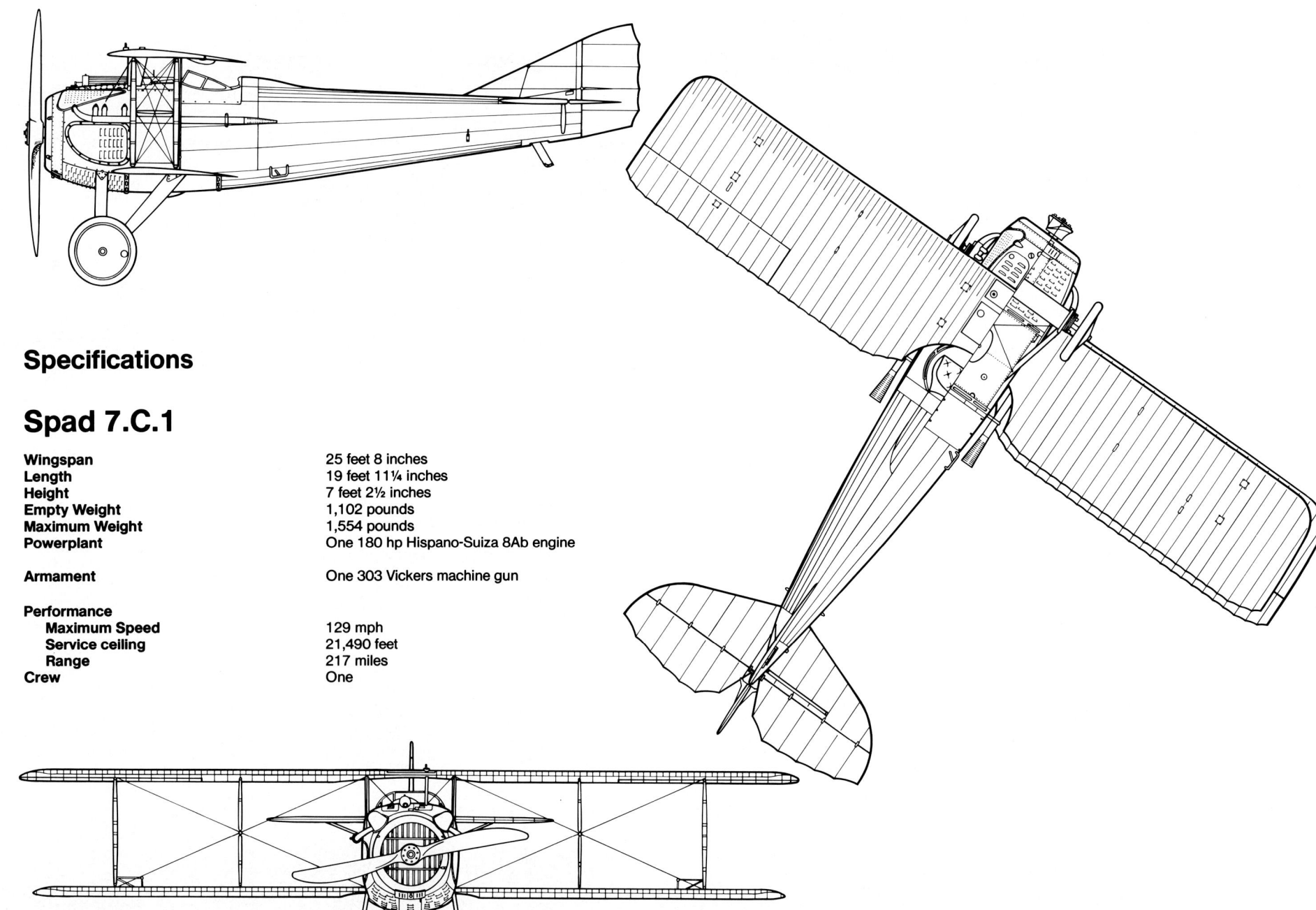

Le Grand Chasseur, LT Georges Guynemer, the best known French ace of the war beside his first Spad 7 (S.113). This was the first of three Spad 7s flown by Guynemer, all of which carried the name *Vieux Charles* on the fuselage. The Stork marking of *Escadrille* Spa.3 was in Red with White details. (Musee de l'Air)

A group of French pilots, including Alfred Heurtaux (Left) and beside him, Guynemer, admire their new Spad 7s. *Escadrille* N3 was the first unit to be fully equipped with Spad 7s during April of 1917. After receiving their Spad 7s the unit was redesignated Spa. 3. (SHAA)

SLT Rene Dorme, called "Papa" by his fellow pilots for the paternal interest he took in them, climbs into his Spad 7. Dorme gained sixteen victories before being killed in combat on 25 May 1917. The round object just below Dorme's foot is the buckle for the pilot's safety harness. (Musee de l'Air)

LT Armand Pinsard beside his Spad 7 named *Revanche IV*. Pinsard brought the first Spad 7 to the front in August of 1916 and scored the first Spad victory on 26 August. *Revanche* means "revenge", or "rematch" — a possible reference to Pinsard's escape from a prisoner of war camp. He scored twenty victories before the war ended. (Musee de l'Air)

This early Spad 7 reveals the single Vickers machine gun, wrap-around windscreen, and early cowling opening. The two teardrop shaped fairings are the covers for the engine cylinder heads and the fairing between the upper wing and the fuselage covered the gravity fuel tank and radiator header tank piping. (Musee de l'Air)

(Above) The small cowling opening of the Spad 7 caused engine cooling problems during warmer weather. Eventually the opening was enlarged, however, before this was done a number of field modifications were tried, including drilling extra cooling holes in the cowling as on this Spad 7. (Musee de l'Air)

(Right) Spad 7s from various *Escadrilles de Chasse* on the airfield at Ham. The Spad in the foreground, is assigned to Spa. 31, while immediately behind it is an aircraft belonging to Spa. 48. (SHAA)

During cold weather, early Spad 7s experienced problems keeping the engines warm enough. A number of field modifications were tried to correct this problem including fitting the radiator with a solid metal cover, such as on this Spad 7. (Musee de l'Air)

These Spad 7s of Spa. 15, parked on a French airfield all carried the plumed knight's helmet insignia on the fuselage side behind the cockpit. The aircraft in the foreground is fitted with the vertical radiator shutters installed on late Spad 7s. (SHAA)

Cowling Development

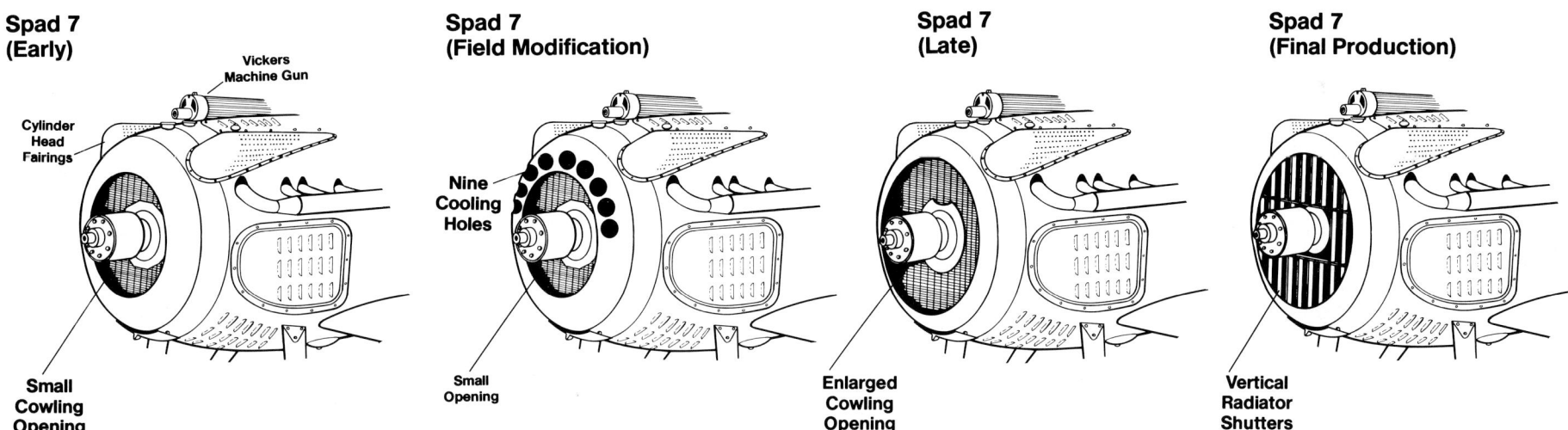

This Spad 7 belonged to an unidentified French pilot who wanted no one to mistake his nationality. The aircraft carries French roundels on the fuselage, wings, and horizontal tail plane. Additionally, he had broad Red, White, and Blue stripes painted on the fuselage sides. The reason for this was to avoid possible confusion with German Albatros DI and DII Fighters. (Musee de l'Air)

This Spad 7 of Spa. 3 at Dunkirk has had the wrap-around windscreen replaced with one from a Nieuport. The Sopwith Triplane in the background may be a Royal Naval Air Service aircraft or it may have belonged to the only French unit that flew the triplane. (Musee de l'Air)

This crashed Spad 7 is believed to have been assigned to *Escadrille* Spa. 67. The aircraft has a pair of eyes painted on the cylinder head fairings. The large square and circular cutouts in the fuselage engine access panels were done to provide additional cooling air to the engine. (Musee de l'Air)

This Spad 7, bearing the Greek warrior insignia of Spa. 31, has a modified windscreen, with the central panel being larger than the standard Spad 7 windscreen. This aircraft has also been fitted with a *Le Chretien* tubular gun sight. (Musee de l'Air)

These Spad 7s of Spa. 23, are believed to have been based at Souilly during July of 1917. The aircraft in the foreground is believed to be a Bleriot-built aircraft. The Black fuselage stripe was repeated across the top surface of the upper wing between the roundels. (SHAA)

Robert Bajac with his Spad 7 named *Nemesis III*. Bajac was assigned to Spa. 48, the unit known as "The Fighting Cocks." This aircraft carries the early, smaller, form of the unit marking, which may have also been repeated on the center section of the upper wing. (SHAA)

It was originally intended that the Spad 7 have a propeller spinner, although this was deleted on production aircraft. This Spad 7 (S.154) of Spa. 62 was fitted with a spinner in the field, believed to have been taken from a Nieuport 17. The spinner is painted Red, White, and Blue in the manner of a French roundel. (Musee de l'Air)

S.154 also carried tri-color stripes on the rear fuselage. The starboard section of the windscreen has been removed, a practice that was often done on the Spad 7 to provide easier access to the gun breech to clear jammed rounds. This aircraft also carries roundels on the tailplane. (Musee de l'Air)

CAPT Georges Guynemer in his third Spad 7 (serial S.254) on 25 May 1917. This aircraft was the first Spad 7 powered by the 180 hp Hispano-Suiza to be sent to the front. Guynemer gained nineteen of his fifty-four victories with this aircraft, which still exists. It was recently restored by the *Musee de l'Air* for display at Le Bourget. (Musee de l'Air)

This Spad 7 (S.239) of Spa. 102 at Courbeaulieu during March of 1917 has the engine cowling access panels removed for maintenance. The fin has been painted a dark color and carries a personal marking, possibly consisting of the pilot's initials. (SHAA)

This Spad 7 (S.1379) was assigned to Spa. 65 and it is believed that the aircraft was assigned to ADJ Marcel Henriot. The diagonal stripes on the fuselage side just behind the winged dragon unit insignia are Red and White. (Musee de l'Air)

CAPT Adrien Leps of Spa. 81 in his Spad 7 named *HOUZARD*. The aircraft also carried Red, White, and Blue stripes painted on the upper wing surface. Leps survived the war with twelve victories. (SHAA)

This Spad 7 was experimentally equipped with a 150 hp Renault engine which required extensive modifications to the aircraft's nose in order to accommodate the 60° angle of the engine's cylinder banks. The engine was not selected for production Spad 7s. (SHAA)

Italian ace Ferruccio Ranza rests next to his Spad 7, which was fitted with a Nieuport-type windscreen. The majority of Spad 7s delivered to the Italians carried clear-doped rear fuselages and Light Yellow nose sections. (Roberto Gentilli)

This Spad 7 carries the Indian head insignia of Spa. 124, the famous *Lafayette Escadrille*. The unit was made up of American volunteer pilots commanded by French officers. (Musee de l'Air)

This Spad 7 is one of approximately fifteen to twenty supplied to the Belgian Air Service and carries the Red comet insignia of the 10e *Escadrille* on the fuselage side. (Musee de l'Air)

An RAF ground crewman demonstrates the use of a Hucks starter truck to start the engine of this Italian Spad 7. The British demonstrated the Hucks starter to the Italians during the mid-1920s. (Roberto Gentilli)

This Spad 7 of an unknown Italian unit carries an inscription on the fuselage side which reads *T vecio...!!!* (old man in the Venetian dialect). The fighter in the background is an Ansaldo S.V.A.5. (Roberto Gentilli)

A trio of Spad 7s of the 77a *Squadriglia* of the Italian *Aeronautica del Regio Escercito* prepare for takeoff from Marcon airfield during the Summer of 1918. The aircraft in the background are Hanriot H.D.1s of the 80a *Squadriglia*. (Roberto Gentilli)

An Italian pilot of the 77a *Squadriglia* alongside the Spad 7 flown by the Italian ace Ernesto Cabruna. The aircraft is fitted with a non-standard windscreen and carries personal markings on the fuselage, and an individual number, XIII, on both the port and starboard upper wing. (Roberto Gentilli)

This Spad 7, fitted with a non-standard windscreen, carries no armament and is believed to have been used as a training aircraft. The marking on the fuselage is believed to be a personal insignia. (SHAA)

Engine Access Panel

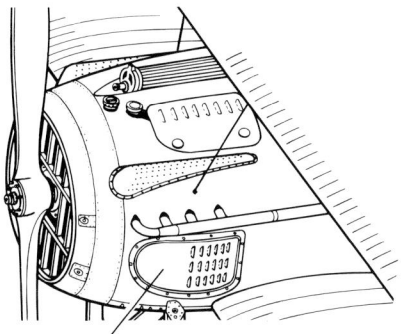

Spad 7 (Standard)

Metal Panel With Eighteen Louvers

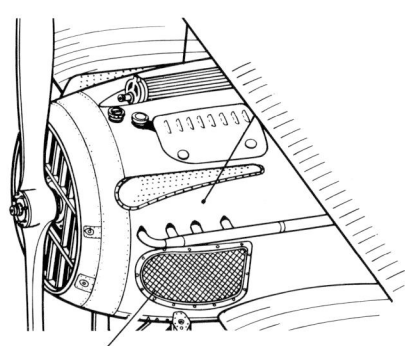

Spad 7 (Modified)

Wire Mesh Panel

LT Rudolf Windisch of *Jasta* 66 in the cockpit of a captured Spad 7. LT Windisch flew the Spad in combat with German markings, however, he retained the dragon marking and tri-color wing stripes of its former owners, *Escadrille* Spa. 65, on the fuselage side. (Musee de l'Air)

This Spad 7 was one of a number of Spads that were exported to Czechoslovakia. The aircraft served with the Czechoslovakian Air Force during 1920 and had the landing gear struts faired over in an attempt to increase the aircraft's top speed. (Zdenek Titz).

This Spad 7, thought to be assigned to S.LT Auguste Ledeuil of *Escadrille* N103 was brought down intact by the Germans on 3 March 1917. The Red star was the early unit marking used by N103. (Hans-Heiri Stapfer)

LT Frederic Loiseau of N 561 in his Spad 7 (S.1068) at Lido airfield. The nose of this aircraft was White and the area around the cockpit is believed to have been Green, with a shield-like marking bearing Loiseau's initials. The charging Black elephant on the fuselage side was also a personal marking. (Musee de l'Air)

During 1976 Guymener's Spad 7 (S.254) was displayed at the *Armee de l'Air* Academy at Salon-de-Provence. The aircraft was hung indoors, however, there was no attempt to restore the Spad until 1981, when it was turned over to the *Musee de l'Air* and completely restored. (Musee de l'Air)

After the war a number of Spad 7s were sold on the civil market and used by civilian pilots. This Spad 7 carries a civil registration on the fuselage side in Black against a White background. (Musee de l'Air)

British Spad 7s

The British Royal Flying Corps (RFC) had expressed an interest in the Spad 7 from the very beginning. Drawings of the prototype, along with reports of its performance, were sent by CAPT Lord Robert Innes-Ker of the British Aeronautical Supply Depot (BASD) to RFC Headquarters, which promptly ordered three Spad 7s from the French.

An example of the new Spad fighter was promised to Innes-Ker in a letter dated 26 July 1916 from the French *Ministre de la Guerre* (Minister of War). The aircraft was to be equipped with an engine taken from the original British order of fifty Hispano-Suizas which had been placed during late 1915. Two additional Spad 7s were promised for later delivery, also fitted with engines taken from British order.

A series of frustrating delays followed, due mainly to the same problems with the Hispano-Suiza engines that the French were having. Finally, on 9 September 1916, *Adjudant* (ADJ) Strohl delivered a Spad 7 (S.126) to No 2 Aircraft Depot (No 2 AD) at Candas, where the aircraft was given the RFC serial A.253. On 27 and 28 September, two additional Spads, serials S.140 and S.143 (later reserialed A.262 and A.263) were delivered.

The first Spad 7 (A.253) was assigned to No 60 Squadron on 20 September for service trials. On 29 September CAPT E. L. Foot flew it on a combat patrol and brought down an enemy aircraft. A.253 was returned to Candas on 17 October and on 23 October it was sent to England.

On 19 September 1916, BGEN R. H. M. Brooke-Popham wrote to the BASD inquiring about the possibility of placing an order for an additional thirty Spads, "...together with a small supply of spare parts, such as landing gear and propellers." Brooke-Popham went on to say that he did not actually want the order placed, but only wished to know if the French would allow the British to place such an order.

CAPT Innes-Ker replied on 23 September that the French would be unable to meet this request with aircraft built by S.P.A.D., however, they advised the British to make arrangements with a sub-contractor, L. Janoir, for the Spads they required. His report also stated that the French were behind in deliveries from S.P.A.D. and could not guarantee the delivery before the end of the year. Nevertheless, the No 2 AD confidently reported on 27 September 1916, that; "It is practically certain that we shall receive from Paris during the last quarter of the year, thirty additional Spads, making a total of thirty-three, and forty-four 150 hp Hispano-Suiza engines."

The order for the thirty Spad 7s was placed in October of 1916, with *Bleriot Aeronautique* instead of Janoir. The first aircraft from this order, serial S.1002 (reserialed A.310), was flown to the No 2 AD on 10 November.

On 30 December 1916, MAJ F. L. Scholte of the BASD wrote RFC Headquarters reporting that "...the supply of Spad machines has been greatly hampered by the weather and by a lack of riggers at Messrs. Bleriot's aerodrome." He also stated that to help speed up deliveries, six RFC riggers had been sent to Bleriot, and it was hoped that nine Spad 7s would be delivered as soon as the weather permitted. It was reported to No 2 AD on 25 January 1917, that ten Spads would be arriving from Le Bourget, with the first scheduled to arrive the following day, however, some of the aircraft would be delivered without guns. By the end of the month, a total of fourteen Spad 7s had been delivered to the RFC.

Despite these delays, the RFC placed an additional order for fifty Spads with the French government. This order was approved by French authorities with the contract going to the firm of *Kellner et ses Fils*. The remaining sixteen Spad 7s of the Bleriot order were finally delivered by 3 March 1917.

No 19 Squadron was the first unit to re-equip with the Spad 7, replacing their earlier

This Spad 7 (S.126) was the first Spad 7 delivered to the Royal Flying Corps at the No 2 Aircraft Depot at Candas. The aircraft was given the RFC serial A.253 which was painted on the fin in Black. (JM Bruce/GS Leslie Collection)

B.E.12s. On 9 October, A.263 arrived, with six additional Spads being delivered during December. The slow pace of deliveries kept No 19 Squadron from becoming fully operational until February, however, by that time, the unit's first victory with the Spad 7 had been scored by LT Capper (22 October).

This Spad 7 (S.211) was delivered to the Royal Naval Air Service in November of 1916 at Dunkirk. The Spad carries the temporary serial N3399 on the rear fuselage in Black. The aircraft was later officially serialed 9611 and was subjected to extensive tests during December of 1916. (JM Bruce/GS Leslie Collection)

No 23 Squadron was the second squadron to re-equip with Spads, replacing the F.E.2b. Their first Spad was received during early February of 1917, with seven more arriving by 24 February. Deliveries remained slow, however, and the unit was not fully re-equipped with Spads until April of 1917.

The Royal Naval Air Service had also shown an interest in the Spad 7 and on 23 November 1916, a Spad 7 (serialed S.211) powered by a 140 hp Hispano-Suiza engine, was delivered to the Naval Station at Dunkirk. This aircraft was reserialed N3399, an identity it bore for a week before receiving its official serial, 9611.

The Spad was tested by Flight Lieutenant A. D. W. Allen at Petite Synthe on 22 December 1916. The top speed was 105 knots at 3,000 feet and climb to 5,000 feet took exactly four minutes. Allen described the maneuverability of the Spad as "...excellent but an inexperienced pilot would find her difficult to turn or throw from one bank to another quickly. Control laterally is good, but not exceptional. A Type 11 Nieuport Scout has more." Allen reported visibility to be; "...good in comparison with the Sopwith Pup, but not as good as the Nieuport."

Before the end of 1916, the British Admiralty began making arrangements to have the Spad 7 built in England by British firms. In late December, orders for fifty aircraft each were placed with Mann Egerton of Norwich, and British Nieuport of Cricklewood. A sample Spad 7 (possibly 9611) was sent to Mann Egerton to serve as a pattern aircraft.

Although it was the Admiralty that ordered the Spad 7 into production in England, no British-built Spad ever served with a Naval Air Service unit. On 14 December 1916, a conference was held to discuss the desperate situation being faced by the RFC in France. Of particular concern was the shortage of available replacement aircraft to replace those lost in combat. At this conference it was agreed to transfer half the Spad 7s under construction for the RNAS to the RFC.

A later conference on 26 February 1917, transferred all the Spad 7s under contract to the RNAS to the RFC in return for the Sopwith Triplanes on order for the RFC. Naval serials (N6800-N6129) had been allotted to the Spad 7s ordered from Mann Egerton, however, these were later changed to N6210-N6284 and N6580-N6604. Only the first British built Spad 7 actually carried a Naval serial. After the order was transferred to the RFC, the naval serials were cancelled and the aircraft were assigned RFC serials A9100-A9161, B1351-B1363, B1364-B1388, and B9911-B9930.

The order for fifty Spad 7s which had been placed with British Nieuport (which had changed its name to Air Navigation Company, Ltd.) was later changed to S.E.5s instead of the Spad 7s. An additional one hundred Spad 7s were also ordered by the RFC from L. Bleriot (Aeronautique) of Brooklands (later Bleriot and Spad). At Bleriot there was a delay in the start of production, because of an incomplete set of production drawings that had been received from the French. To correct the problem, a single pattern aircraft (possibly A.253) was sent to Bleriot to serve as the basis for new production drawings.

The first Mann Egerton Spad 7, still carrying its RNAS serial N6210, was sent to Martlesham Heath for official trials in April of 1917. This aircraft had provisions for mounting a Lewis machine gun on the upper wing center section (although only the fuselage-mounted Vickers gun was carried during the early flight tests). The trials revealed a top speed of 119 mph at 6,500 feet, with a time to climb of six minutes, twenty-one seconds. A handwritten note on the test report summary claimed that the results were, "...practically the same as the French Spad..."

One of the major concerns of the RFC was the Spad 7's light armament and tests were conducted with various installations of a second machine gun on the Spad 7. The first Spad 7 (N6210, reserialed A9100) was tested at Martlesham Heath in May of 1917, with a upper wing mounted Lewis gun, however, the installation caused a slight drop in performance. Top speed was reduced by some three mph and the rate of climb was cut by one and one half to three minutes. According to a letter from Brooke-Popham to the 5th

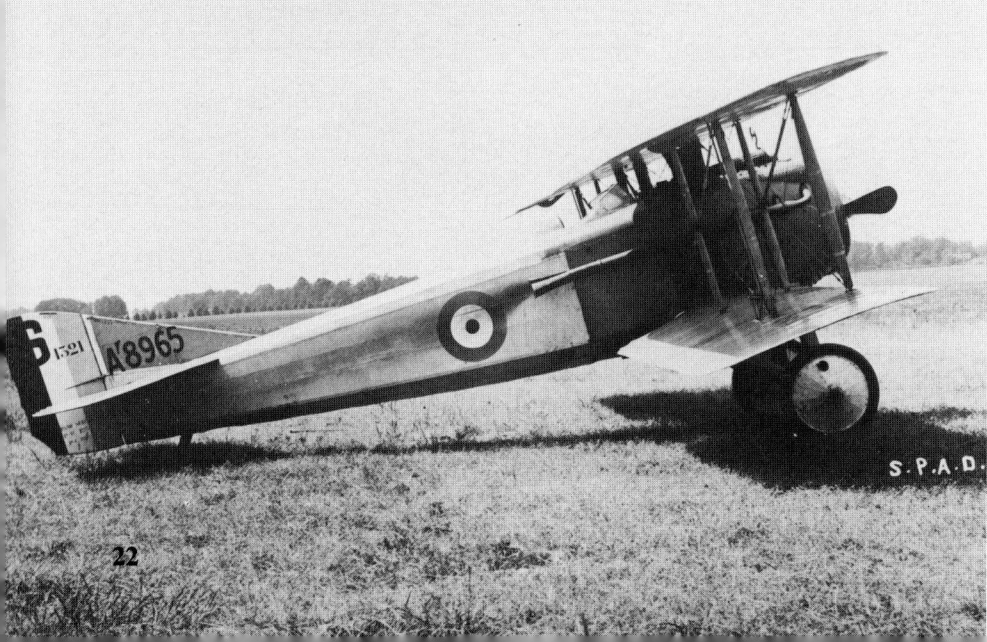

This Spad 7 (S.1321) was originally supplied to the Royal Naval Air Service by the French. During April of 1917 it was turned over to the RFC and reserialed as A8965, which was painted on the fin in Black. (IWM)

This Spad 7 (S.1027) was part of the original contract for thirty Spad 7s ordered from *Bleriot Aeronautique* by the British. It was given the RFC serial A6714 and assigned to No 19 Squadron at Vert Garand during April of 1917. (RAF Museum)

Brigade, RFC, dated 20 September 1917, an attempt was made to mount a Lewis machine gun to the upper wing of a No 23 Squadron Spad 7. The field installed mount was similar to that used on the S.E.5 (it is believed that the installation consisted of a Foster sliding rail gun mount modified from an S.E. 5 mount). It was reported that this arrangement also reduced the Spad's performance and it was not adopted for widespread use.

One of the problems the British had with French-built Spads also concerned the aircraft's armament. Early Spad 7s had the ammunition wound around two drums; one mounted in front of the cockpit which fed the cartridges to the gun and another mounted behind the cockpit which wound up the empty cartridge belt. Double ammunition feeds often occurred when the momentum of the feed drum forced the ammunition belt onward after the pilot had ceased firing. After the Prideaux disintegrating-link ammunition belt became available during November of 1916, an ammunition box was installed replacing the belt drums thus ending the feed problem.

British Bleriot-built Spad 7s featured a fairing in front of the cockpit over the breech of the Vickers machine gun which inhibited the pilot's forward vision, particularly for sighting. In a letter dated 27 May 1917, Brooke-Popham ordered the fairing removed as soon as the aircraft arrived at the No 1 Air Depot in France. The fairing was retained on Spad 7s assigned to training units in England and on aircraft assigned to the Middle East.

As a general rule, British-built Spad 7s were not as good as the French-built aircraft. Their performance was poor and they were not as popular with British pilots. In a letter dated 28 July 1917, Brooke-Popham wrote: "The performance of English-built Spads is greatly inferior to that of the French-built Spads. The air speed indicator readings of French-built Spads flying level at a low altitude varies from 120 to 128 (mph), that of the English-built Spads from 108 to 110. The climb and ceiling are also not as good, and in addition the English-built Spads are nose heavy."

Some of the complaints registered against British-built Spads included the fabric on British Spads not sewn to the wings properly, tail skids of an inferior wood, and radiators not up to the French standards. Because of their inferior performance in comparison with the French-built Spad 7s, very few, if any, Mann Egerton built Spads were sent to France. Most were relegated to training duties in Britain. In the event, the British preferred to equip their operational Spad squadrons with French-built machines. On 28 July 1917, Brooke-Popham reported that; "So far as can be foreseen at present, the supply of Spads from French sources will be sufficient to keep up our two squadrons until the end of the year without drawing on the English supply."

Production of Spad 7s in England was halted after production of a total of 220 aircraft — 120 by Mann Egerton and 100 by British Bleriot. The French supplied a total of 185 Spad 7s.

No 19 Squadron began to replace its Spad 7s with Sopwith Dolphins during November of 1917 with No 23 Squadron relinquishing their Spad 7s for Spad 13s the following month. The Spad 7 continued to serve with training units in Britain, along with Nos 30 and 63 Squadrons in Mesopotamia, and No 72 Squadron in Palestine.

This Bleriot-built Spad 7 was used by No 23 Squadron and was painted Light Grey overall. The radiator cowling and fuselage stripes are believed to have been Red and White. The fin may also have been painted Red. (RAF Museum)

This Spad 7 (B.6772) is believed to have been assigned to No 23 Squadron. The Spad 7 is camouflaged in the standard British finish of Khaki Brown (P.C. 10) uppersurfaces over clear doped undersurfaces. The metal panels on the nose were painted Light Grey. (Musee de l'Air)

This Spad 7 (N6210), built in England by the Mann Egerton Company, was the first aircraft off the production line. It carries the company name and address under the horizontal stabilizer in White. (RAF Museum)

Early British built Spad 7s featured small radiator cowling openings that were later enlarged to overcome cooling problems. The British built Spad 7s were inferior in a number of respects to the French aircraft and were used mainly as trainers in England. (IWM)

Experimental Upper Wing Gun Mount

British-built Spad 7

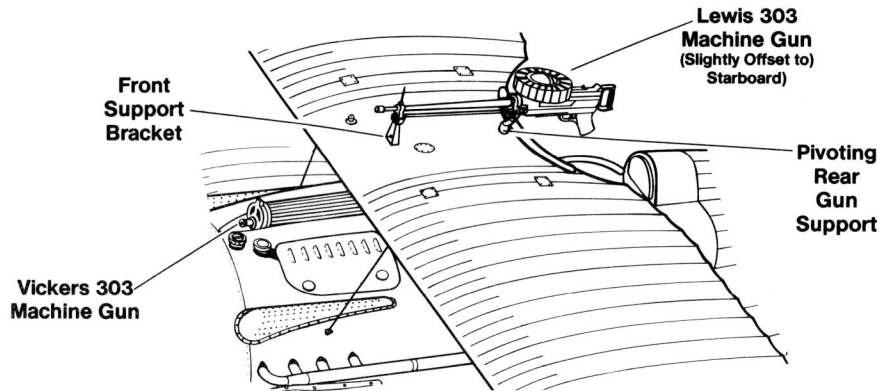

This Spad 7 (A9100), formerly serialed N6210, was the first Mann-Egerton Spad 7, and was tested at Martlesham Heath in May of 1917 with a 303 Lewis machine gun mounted on the starboard side of the upper wing center section. Tests revealed that the gun installation reduced the aircraft's performance slightly. (IWM)

A ground crewman spins the propeller on a British Spad 7, while two other crewmen hold down the tail. This Spad 7 (A 8799) was built by the British Bleriot Company and is believed to have been assigned to a training unit based in England. (RAF Museum)

Cowling Development

French-built Spad 7

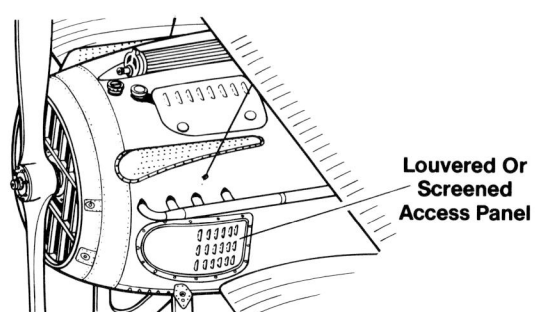

Louvered Or Screened Access Panel

British Bleriot-built Spad 7

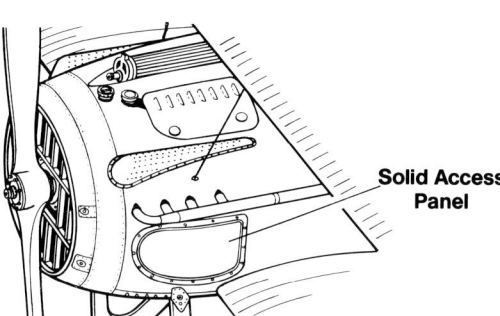

Solid Access Panel

A British Bleriot-built Spad 7 (A8878) shares the field with an unarmed Sopwith Pup. Both aircraft are believed to have been assigned to a training unit in England. (RAF Museum)

This Spad 7 (A 8798) was also built by British Bleriot. One of the key indentification features of the British Bleriot Spad 7 is the absence of louvers on the forward fuselage access panel. This aircraft has had the gun breech fairing removed. (IWM)

This unarmed British Bleriot Spad 7 served with No 2 Fighting School, at Marske during 1918. The aircraft has been equipped with oversized tires. The different colors of the engine panels suggests that they were probably cannibalized from another Spad 7. (RAF Museum)

This British Bleriot-built Spad 7 undergoes maintenance at a Royal Flying Corps airfield. The engine access panel on the port side of the nose has been removed to allow access to the engine. (RAF Museum)

A British pilot in the cockpit of his Spad 7. British Bleriot-built Spad 7s were fitted with a hood-like fairing over the breech of the Vickers gun. This fairing blocked the pilot's forward vision and was ordered removed on aircraft sent to France. (RAF Museum)

Machine Gun Breech Fairing

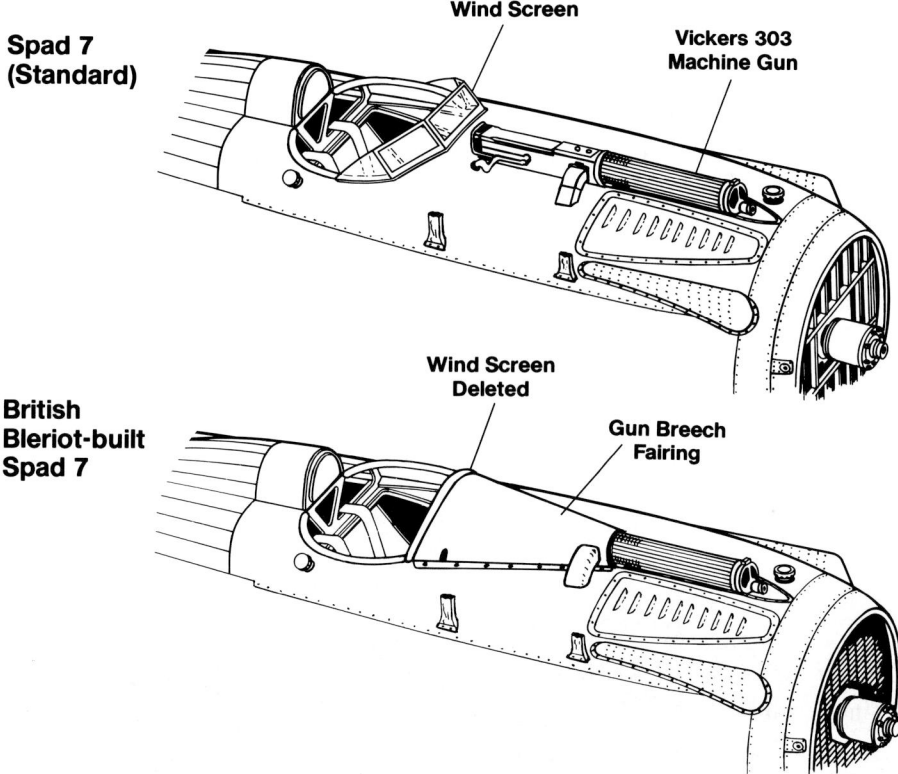

(Above) This British Bleriot-built Spad 7 (A8817) at London Colney airfield has had the top part of the cowling cut away exposing the engine cylinder banks. The aircraft has also been fitted with a small spinner on the propeller hub and appears to have a camouflaged nose.

(Right) This Spad 7 of No 30 Squadron was based at Baqubah during 1918 and carries a Lewis machine gun on a mounting designed by CAPT Hereward de Havilland. The tubular object ahead of the pilot is an Aldis gun sight. A small windscreen is fitted above the gun breech hood. (RAF Museum)

Ground crewmen guide the pilot of this British Spad 7 out for takeoff. This Spad 7 (A8797) was the fourth British Bleriot-built Spad 7 and was probably used by a training unit. This aircraft has had the fairing in front of the cockpit removed and replaced with a modified windscreen. (RAF Museum)

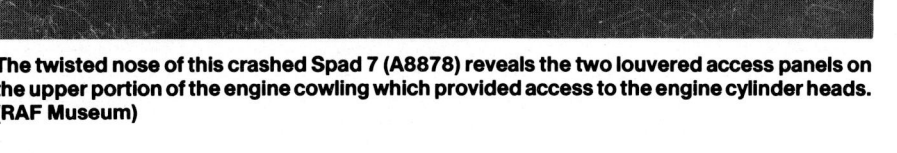

The twisted nose of this crashed Spad 7 (A8878) reveals the two louvered access panels on the upper portion of the engine cowling which provided access to the engine cylinder heads. (RAF Museum)

This British Bleriot-built Spad 7 (A8812) of No 30 Squadron at Baqubah in the Middle East during 1918 has been fitted with a spinner and is armed with a Lewis gun mounted on top of the upper wing center section. (RAF Museum)

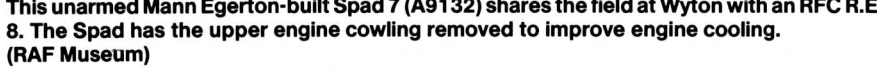

This unarmed Mann Egerton-built Spad 7 (A9132) shares the field at Wyton with an RFC R.E. 8. The Spad has the upper engine cowling removed to improve engine cooling. (RAF Museum)

A pair of Mann Egerton-built Spad 7s (A9150 and A9159) parked outside one of the hangars at London Colney airfield. The last two digits of the Black serial number on the rudder of the Spad 7 in the foreground are outlined in White. (RAF Museum)

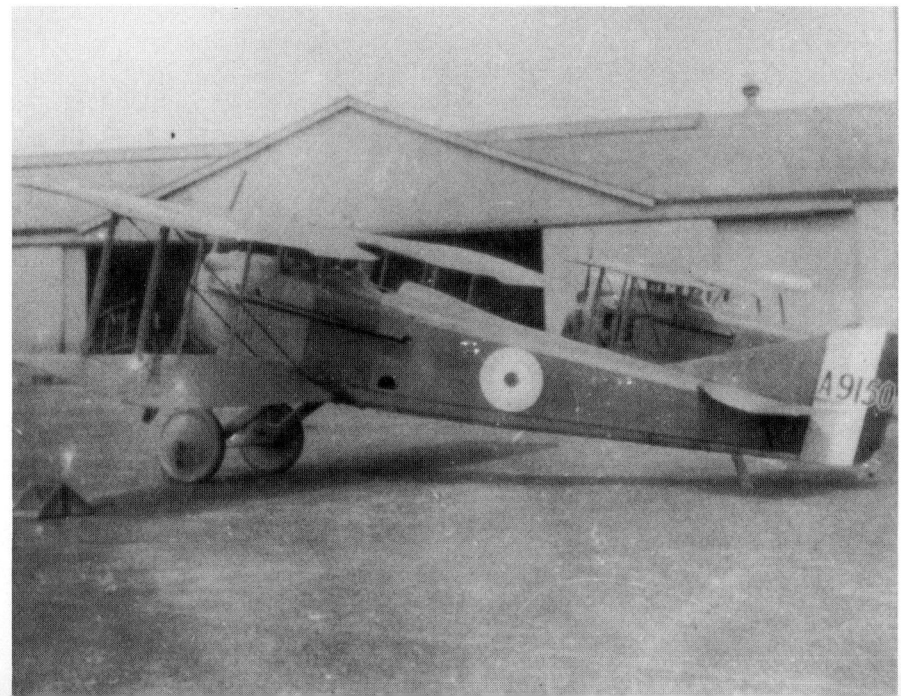

An RFC pilot walks out to an unarmed Spad 7 assigned to a training unit in England. The interplane struts have been painted White and a White stripe has been added to either side of the fuselage roundel. (RAF Museum)

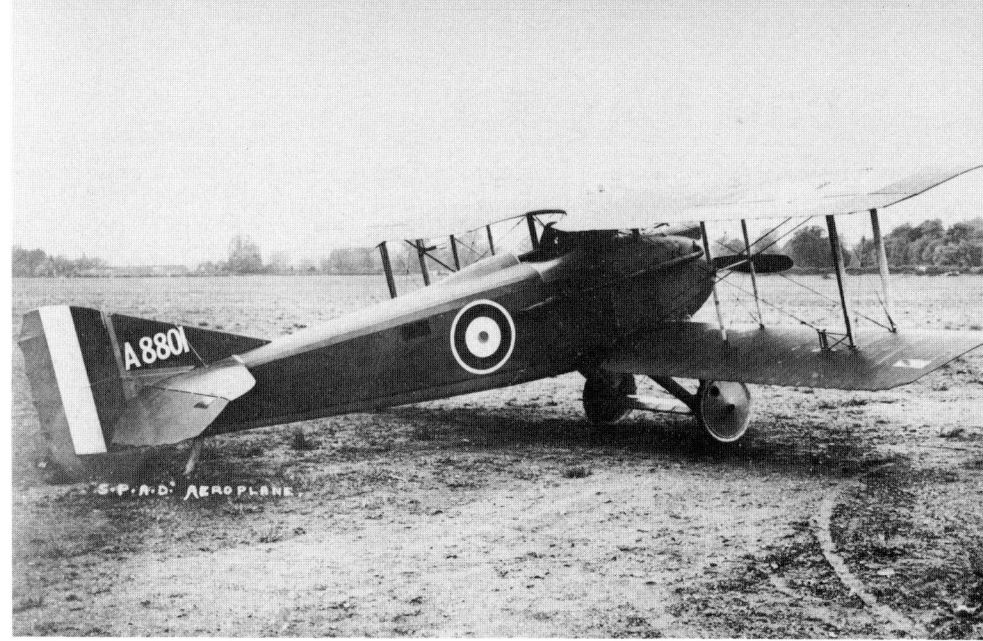

One of the differences between British Bleriot-built Spad 7s and Mann Egerton-built Spad 7s was in the manner of the presentation of the fuselage roundel. British Bleriot Spads had the fuselage roundel further forward than on Mann Egerton-built Spads. (IWM)

This unarmed British Bleriot Spad 7 of a training unit in England, has a White stripe painted on the fuselage side. The cylinder bank fairings have been removed and the top of the cowling cut away to improve engine cooling. (RAF Museum)

While a pair of ground crewmen hold down the tail, a third crewman prepares to spin the propeller to start this British Bleriot Spad 7. The two crewmen on either side will pull out the wheel chocks once the Spad is ready to taxi out. (RAF Museum)

Spad 12.Ca.1

During late 1916, Georges Guynemer approached S.P.A.D.'s chief designer, Louis Bechereau with the idea of arming a Spad fighter with a cannon mounted to fire through the hollow propeller shaft of a geared 200 hp Hispano-Suiza. The idea appealed to Bechereau and he began work on the cannon-armed Spad under the designation Spad 12.Ca.1. The first official mention of the Spad 12 was in official French reports dated 10 December 1916.

The Spad 12 was very similar in appearance to the earlier Spad 7, however, it was larger and heavier. The Spad 12 was powered by a 200 hp Hispano-Suiza 8C engine, with a short-barreled 37MM Hotchkiss shell-firing cannon mounted between the cylinder banks, firing through the propeller hub. The engine cowling was more streamlined than the Spad 7, with the characteristic teardrop-shaped cylinder bank fairings being eliminated. Besides the cannon, the Spad 12 also carried a synchronized Vickers machine gun, mounted on the upper fuselage decking offset to starboard in front of the cockpit.

The wing area of the Spad 12 was increased from 17.85 square meters (192.137 square feet) of the Spad 7 to 20.20 square meters (217.43 square feet) and the wings had a slight positive stagger and rounded wing tips. The rudder and elevator trailing edges were also more rounded than those of the Spad 7.

The Spad 12 was considerably heavier than its predecessor, with a loaded weight of 890 kg (1,948 pounds) — 185 kg (407 pounds) more than the Spad 7. The Hotchkiss cannon accounted for a good part of this increase, weighing 45 kg (99.208 pounds) with an additional 10 kg (22.046 pounds) being added for the twelve shells supplied for the cannon.

The cockpit of the Spad 12 was deep and fitted with a small windscreen which, on the prototype, featured a generously padded forward cockpit coaming. The breech of the Hotchkiss cannon extended into the cockpit between the pilot's knees, making it impossible to fit a traditional control column to the cockpit. Instead, a Deperdussin-style control system, similar to those used in pre-war monoplanes, was used to operate the elevators and ailerons. These controls were recommended by Guynemer in a letter to Bechereau dated 8 January 1917.

Guynemer personally carried out much of the testing of the Spad 12 during the spring of 1917. Test reports give the top speed as approximately 220 kph (137 mph) at ground level and 195 kph (121 mph) at 5,000 meters (16,400 feet), time to climb to 5,000 meters was approximately twenty minutes, and the aircraft's ceiling was 7,000 meters (23,000 feet).

Guynemer had his Spad 12 (S.382) at the front by July of 1917 and, on the fifth of that month, he flew his first combat sortie. He attacked three D.F.W, two seat observation aircraft; however, before he could score a kill, the Spad 12 was damaged by the Germans' return fire, forcing him to land. S.382 had to be sent to Buc for repairs, which took almost three weeks. On 27 July 1917, Guynemer scored his first victory in the Spad 12. With one shell from the Hotchkiss cannon and eight rounds from the Vickers machine gun, he destroyed an Albatros two-seater. The next day he shot down a D.F.W. with two cannon shells and thirty machine gun rounds. Unfortunately, S.382 was again hit by enemy fire and seriously damaged, forcing it out of action until mid-August.

Based on the success of the tests and Guynemer's victories, the Spad 12 was ordered into production. The initial contract called for 300 aircraft, however, it is doubtful that production ever reached this number, since there were never a great number of Spad 12s in service at any given time. It is believed that one or two Spad 12s were assigned to various *Escadrilles de Chasse* with instructions to let each unit's best pilots use the aircraft.

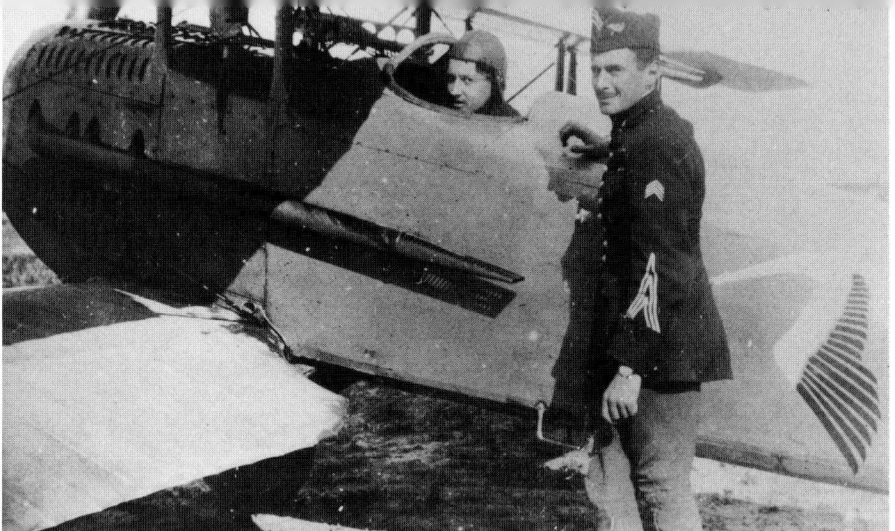

Georges Guynemer in the deep cockpit of the first Spad 12 (S.382). Guynemer personally conducted much of the flight testing on the Spad 12. The bracket on the center-section struts was used to hold a camera. During testing, the aircraft carried the Stork insignia, but none of Guynemer's personal markings. This Spad 12 has been fitted with a padded windscreen coaming. (Musee de l'Air)

The Spad 12 was reportedly difficult to fly, even for the most experienced pilots. Using the cannon in combat required a good deal of skill, primarily because of the difficulties in aiming the weapon. Some pilots used the Vickers gun as a sighting weapon, firing the cannon when they saw the machine gun's tracers hitting the target. The cannon also had a heavy recoil and fumes from the breech filled the cockpit after each firing (it is possible that the Spad 12 did not have a firewall, since contemporary drawings of the internal arrangement do not show one.) Another difficulty was that the Hotchkiss cannon fired one round at a time and reloading it in combat was extremely difficult.

Despite these problems, the Spad 12 was used by several well known French aces, including Rene Fonck who flew two Spad 12s (S.445 and S.452). He gained eleven of his

After the Spad 12 was delivered to Guynemer, he added the name *Vieux Charles* and a diagonal White stripe to the fuselage sides. The individual aircraft number, 2, was in Black. (JM Bruce/GS Leslie Collection)

seventy-five victories with the Spad 12. Fourth-ranking French ace Georges Madon flew a Red painted Spad 12 with *Escadrille* Spa 38. Others who successfully used the Spad 12 included Albert Deullin, Fernand Chavannes, and Lionel de Marmier. LT Francois Battesti flew a Spad 12 with Spa 73 during October of 1918. He was reportedly pleased with the aircraft and is believed to have gained his seventh victory with it on 29 October.

Guynemer seemed to prefer the heavily armed Spad 12 and gained four victories (his 49th - 52nd) in S.382. Unfortunately, the aircraft was under repair most of the time it was assigned to Guynemer.

Two Spad 12s were exported. One, serial S.449, was delivered to the Royal Flying Corps in January of 1918. This aircraft was reported to be at Buc on 5 January, and was delivered to the No 2 Aircraft Depot by ADJ de Courcelles on 9 March. The Spad 12 was given the RFC serial B6877 and was flown to England on 18 March, where it was tested at Martlesham Heath. One RFC pilot reported that the Spad 12 was a "...soggy, nose-heavy job..." — an assessment shared by many of the French pilots who flew the Spad 12 operationally. B6877 was sent to the Isle of Grain on 4 April 1918, but crashed enroute. The aircraft was not rebuilt and no replacement was ordered.

At least one Spad 12 was taken into service with the United States Air Service. Ordered in July of 1918, it was originally intended to go to the 139th Aero Squadron, where it was to be flown by LT David E. Putnam. Putnam was killed on 14 September 1918, before the aircraft was delivered and it was passed to the 13th Aero Squadron, where it was flown by the unit's CO, MAJ Charles J. Biddle.

Guynemer displays his *avion magique* to General Francet d'Esperey at La Bonne Maison on 5 July 1917. Earlier that day, Guynemer had flown the first combat mission with the Spad 12, during which the aircraft was damaged. The wings of S.382 have been removed, pending its transport to Buc for repairs. (IWM)

This early Spad 12 reveals the revised cowling which lacked the prominent cylinder head fairings and repositioned exhaust stacks which were lower on the fuselage sides than on the earlier Spad 7. This aircraft was a test machine and had the fuselage undersurfaces painted Black. (SHAA)

Cowlings

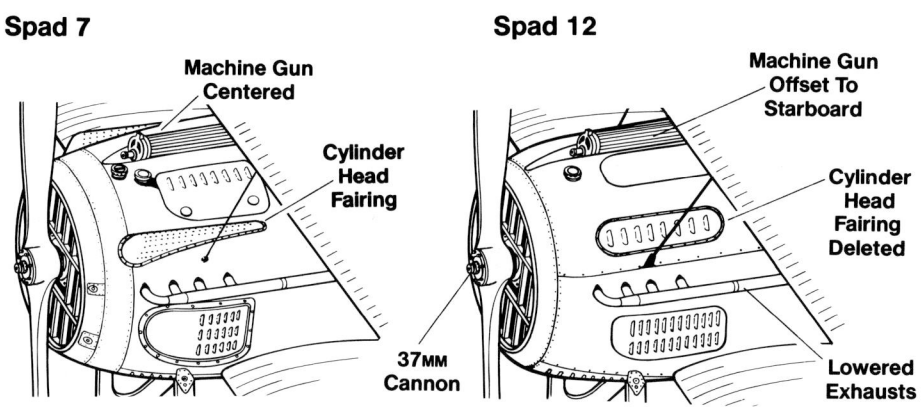

An early Spad 12 (S.434) parked on a French airfield. Most units had one or two Spad 12s assigned to them, with instructions to allow the best pilots in the unit to fly the aircraft. This Spad 12 is believed to have been painted overall Light Grey. (NASM)

Two Spad 12s (foreground) share the field with five Spad 7s, at a *Reserve Generale de l'Aviation* (RGA) airfield while awaiting delivery to French *Escadrilles de Chasse*. The two aircraft carry different color schemes; the aircraft on the left is painted an overall dark color, while the aircraft on the right has been camouflaged. (Musee de l'Air)

Spad 12s were progressively modified during production with new engines and other improvements. This late Spad 12 is powered by a 220 hp Hispano-Suiza 8Bc engine and has the louvered engine access panels removed for maintenance. (Musee de l'Air)

Wing Stagger

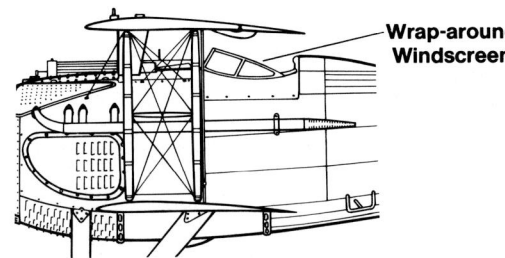

**Spad 7
No Wing
Stagger** — Wrap-around Windscreen

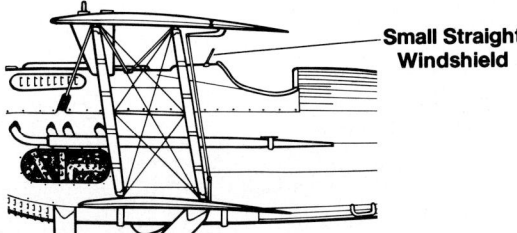

**Spad 12
Slight Positive
Wing Stagger** — Small Straight Windshield

This camouflaged late production Spad 12 has been modified with squared off wings tips. The Vickers machine gun on the Spad 12 was offset much further to starboard than the gun on the earlier Spad 7. (Musee de l'Air)

CAPT Rene Fonck, France's leading ace, alongside one of two Spad 12s (S.452) he flew with Spa. 103. Fonck scored eleven of his seventy-five victories with the Spad 12. Although arrogant and boastful, Fonck was a superb pilot and marksman who, on two occasions, shot down six German aircraft in a single day. (Musee de l'Air)

This Spad 12 (S.440) reveals the plywood pocket extensions that were fitted to the wing tip trailing edges of Spad 12s to square off the wings. Later, new wings incorporating square tips were introduced onto the production line. (IWM)

Wing Tips

Spad 12 (Early) — Rounded Tips

Spad 12 (Late) — Plywood (Pockets) Extensions

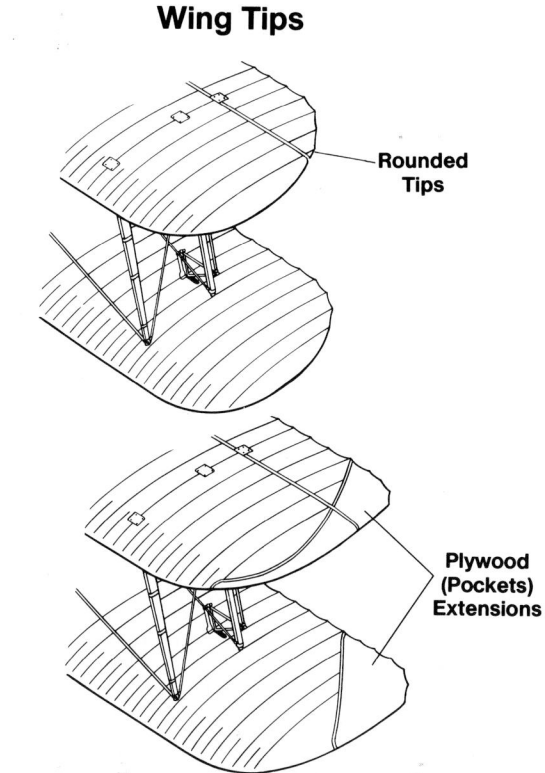

(Above) The other Spad 12 assigned to Fonck was S.445. Fonck's personal number, VI, was carried on the rear fuselage and repeated on the port upper wing surface in Red, outlined in White. The White-outlined Red star on the starboard wing was the former unit insignia of Spa.103. (Musee de l'Air)

The United States Air Service acquired one Spad 12 which was flown by MAJ Charles J. Biddle, commander of the 13th Aero Squadron. The aircraft was marked with a White outlined Black O on the fuselage side and was modified with a circular clear inspection panel on the lower fuselage side, between the exhaust pipe and the wing root. (JM Bruce/GS Leslie Collection)

(Below) Fourth-ranking French ace Georges Madon in the cockpit of the Spad 12 he flew with Spa. 38. The fuselage and tail of this aircraft were painted Red, with the wings left in the standard clear-dope finish applied to early Spad 12s. (SHAA)

Spad 13.C.1

By the Spring of 1917, the Spad 7 was rapidly being overtaken by the current generation of twin gun German fighters; Bechereau began searching for ways to improve the Spad 7. Earlier, the Hispano-Suiza engine had been improved and by May of 1916, Marc Birkigt had developed a geared 200 hp version under the designation 8Ba. With the availability of this engine, Bechereau began development of a new fighter under the designation Spad 13.C.1

The earliest known reference to the existence of the Spad 13 is found in a letter from MAJ F. N. Scholte, of the British Aviation Commission, to RFC Headquarters, dated 20 February 1917; "For your information, Messrs. Spad (sic) have now on order for the French Government twenty 200 hp (with geared down propeller) Hispano-Suiza single seater Spads which are to be fitted with two Vickers guns, mounted upon the cowling in the usual manner and firing through the propeller to the right and left, respectively, of its axis." Another letter from MAJ Scholte, dated 13 March 1917, further described the new Spad and reported that; "...the first machine is now ready and awaiting favourable weather for trials."

This aircraft, Spad 13 (S.512), was being subjected to extensive tests at Villacoublay. It is believed that instead of constructing a single prototype, the French found it more expedient to build a pre-production batch of twenty Spad 13s. One aircraft was kept for performance trials and the remainder were sent to the front for combat trials.

The Spad 13 bore a strong family resemblance to its predecessors, the Spad 7 and Spad 12, however, it was slightly larger. The fuselage was five inches longer than the Spad 7 and the wing span was one foot five inches longer. The Spad 13 was also heavier weighing 1,888.5 pounds, compared with 1,544.5 pounds for the Spad 7. The cabane struts were given a slight forward stagger, although the wings were rigged with no stagger. The Spad 13 wing differed from the Spad 7 wing in having rounded wingtips, similar to early Spad 12s. The engine cowling was more tapered than that of the Spad 12, and the cylinder bank teardrop fairings were re-introduced, although these were not as prominent as those on the Spad 7.

The Spad 13's primary improvement over the Spad 7 was in its armament. The Spad 13 carried double the fire power of the Spad 7 with two Vickers 303 machine guns being mounted in deep troughs on the upper fuselage decking in front of the cockpit. Ammunition boxes were located in the fuselage between the guns holding a total of 800 rounds of ammunition (400 rounds per gun). The guns had individual triggers and could be fired either separately or together. Spent shell casings were ejected into a chute that funneled them away from the pilot and out through a port on the lower fuselage just above the wing.

Test flights with S.512 began on, or about, 22 March 1917 with LT Salze at the controls. Tests revealed a top speed of 211 kph (131 mph) at 1,000 meters (3,280 feet), and 190 kph (118 mph) at 5,000 meters (16,400 feet). Time to climb was two minutes twenty seconds, and twenty minutes ten seconds, respectively. The aircraft's theoretical ceiling was given as 6,800 meters (22,300 feet), however, it is not known if the Spad 13 actually achieved this altitude.

By 4 April 1917, another Spad 13 (S.392) was being flown at Buc by S.LT Rene Dorme, while Maurice Prevost was demonstrating another to the *Escadrilles de Chasse*. The RFC ordered CAPT C. K. C. Patrick to visit the aerodrome at Fisme on 29 April and to inspect the new Spad. He was not to fly the machine, or even ask to fly it. He was merely to report on visibility, the amount of ammunition carried, the accessibility of the guns in the event of a jam, and "...any points regarding the fittings on the machine." He was also to ascertain, from the French test pilots, details of the aircraft's performance, and their opinions of its maneuverability.

Patrick reported that the view from the cockpit was "...practically identical with the 150 hp S.P.A.D. (sic)." He also reported that the Spad 13 carried four hundred rounds of ammunition for each gun. As for being able to clear jams, Patrick reported; "The right hand gun is in much the same position for getting at as the gun on the 150 hp S.P.A.D. (sic). Without moving the windscreen and sights, it would be impossible to remedy any but the simplest jambs (sic) on the left hand gun." Patrick also reported that the speed of the Spad 13 was 120 mph at 4,000 meters (13,120 feet), with a time to climb of eleven minutes. "The machine is said to be less handy near the ground," Patrick went on, "but considerably handier at height than the 150 hp S.P.A.D."

Based on these tests, the French Government ordered the Spad 13 into production, with an initial contract for 250 aircraft. Unfortunately, as with the earlier Spad 7, early production was slow, due primarily to teething troubles with the 200 hp Hispano-Suiza engine and other technical problems.

Some of these problems and possible ways of solving them, were discussed at a meeting of the *Section Technique de l'Aeronautique* (Aeronautical Technical Section, STAe) on 3 August 1917. One problem concerned the main fuel tank, which was bulging due to overpressure. This was apparently caused by a malfunctioning pressure relief valve. Recommendations were made to install a different type of valve and a gauge to enable the pilot to monitor the fuel tank pressure.

Displacement of fuel to the front and rear of the main fuel tank during dives and climbs (when fuel levels were below 30 to 35 liters) caused interruption of fuel to the engine. The solution recommended was to partition the fuel tank into two compartments, each with its own feeder pipe. The tank was also subject to buffeting, for which a strengthened mounting was suggested. LT Salze, the test pilot who had flown S.512 during its tests at Villacoublay, reported that the rudder response was inadequate and Bechereau proposed installing a larger fin to eliminate the problem.

Recommendations were also made to improve the visibility for the pilot by revising the aircraft's cowling and to improve the control arrangement by repositioning the control cables running under the pilot's seat.

This Spad 13 (S.512), one of the first twenty pre-production aircraft built, was used for official tests by the *Section Technique de l'Aeronautique* at Villacoubly from March through September of 1917. (Musee de l'Air)

Radiators on the Spad 13 were also a problem and were the subject of another conference at the STAe on 25 August 1917. This conference was called because an "...inadmissable proportion..." of Spad aircraft were unserviceable due to radiator leaks.

These leaks were caused by the way the radiators were mounted in the aircraft. The Spad 13 airframe had been designed to accept more powerful and larger versions of the Hispano-Suiza engine, making it necessary to mount the radiator so it could be moved fore and aft. The radiator was not attached to the fuselage structure, but rather it was simply bolted to the front of the engine. Engine vibration (a common problem with the 200 hp Hispano-Suiza engine) and the shock of rough landings, caused leaks to develop around the attachment lugs, behind the central support angles, behind the vertical joining plane of the radiator tubes, or in the blades of the central grill. Various ways of solving this problem for both existing aircraft and those under construction were discussed.

All radiators returning from the front for repair were to be repaired by their respective manufacturers, and were to be modified with a number of reinforcements. The central tank was to be changed to one made of 1MM sheet metal, the attachment lugs were to be fixed to the grill without connection to the radiator core, and the surface area of the lugs was to be doubled.

The radiator support mounting was also modified with the aluminum support bands of the exterior cowling being replaced with bands made of 1MM sheet steel, the screen which covered the radiator rim was to be mounted with washers placed between the screen and the support structure, and the radiator was to be fixed in place only by the eyebolts of the cowling, allowing it to move freely around its central support.

These modifications were to be made in workshops in the interior and at front line squadrons according to a notice, with sketches drawn up at the S.P.A.D. works, in accordance with the instruction issued by the STAe.

All of these improvements and modifications were completed or were being carried out by 20 November 1917. Still other improvements were recommended at that time, including a revised wing which replaced the rounded wing tips with squared off wing tips to improve handling. Before the new wings were incorporated into the production line, a field modification was devised in the form of plywood "pockets" which were to be fitted to the outer edges of the wing tips, as had been done with the earlier Spad 12.

Another modification was the fitting of a wooden fairing to the drag wire that ran from the top of the forward cabane strut to the top of the fuselage. This fairing gave the drag wire the appearance of an additional strut. In order to further improve engine cooling, it was recommended that the louvered access panels on the fuselage sides be replaced with screened panels.

Problems continued to hamper deliveries of the Spad 13 and by 21 December 1917, the unavailability rate was still between 40 and 60 percent. The engine mounts caused some of these problems because, like those of the earlier Spad 7, they were not strong enough to stand up to the engine vibration. This was solved by reinforcing the mount with two lateral steel plates. Another problem was with the engine lubrication system; oil was not getting to the engine, causing severe engine damage. To fix the problem a larger diameter feed pipe was installed between the oil pump and the oil tank to facilitate the movement of oil.

Testing of the Spad 13 (S.512) continued at Villacoublay during September with ADJ Bourgeois at the controls. In one series of tests, the aircraft was flown with a modified 220 hp 8Bec engine and three different propellers. Tests revealed that the Ratmanoff type 6727 propeller gave the best results, with a speed at 2,000 meters (6,560 feet) of 218 kph (136 mph) and 203 kph (126 mph) at 5,000 meters (16,400 feet). Time to climb was four minutes forty seconds and eighteen minutes thirty seconds, respectively.

As the technical problems with the Spad 13 were resolved, production accelerated during 1918, with production being shared by S.P.A.D. and eight subcontractors. The total number of Spad 13s built is unknown, however, estimates run between 7,300 and 8,472. The former figure was cited by Commandant Georges Guignard (head of the *Section Fabrication de l' Aeronautique*) and is believed to be the most accurate. The latter figure is the production total of all firms building the Spad 13, which may represent the number of aircraft ordered, rather than actually produced.

This early Spad 13 is believed to have been one of the twenty pre-production aircraft built during early 1917. These aircraft had a more tapered engine cowling than the Spad 12 and also reintroduced the teardrop-shaped cylinder bank fairings, similar to those used on the Spad 7. (Musee de l'Air)

Cowling And Armament

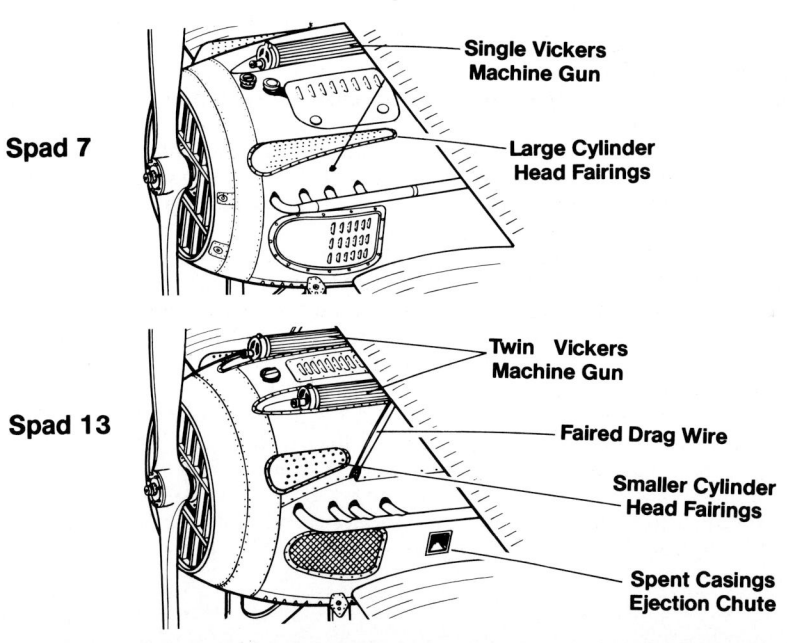

Spad 7
- Single Vickers Machine Gun
- Large Cylinder Head Fairings

Spad 13
- Twin Vickers Machine Gun
- Faired Drag Wire
- Smaller Cylinder Head Fairings
- Spent Casings Ejection Chute

S.P.A.D.	1,141
A.C.M. des Colombes	361
Adolphe Bernard	1,750
Bleriot Aeronautique	2,300
Societe Anonyme des Etablissements Borel	300
Avionnerie Kellner et ses Fils	1,280
Pierre Levasseur	340
Societe Anonyme des Etablissements Nieuport	700
Societe Anonyme Francais de Constructions Aeronautique	300

The British were naturally interested in the Spad 13 and had tested an early example (serial S.489) at No 2 Aircraft Depot during early June of 1917. Given the RFC serial B3479, the aircraft was test flown briefly on 6 June, but was forced to land because of engine over-heating. After some adjustments, a full battery of tests was carried out the following day, during which the aircraft showed a top speed of 135 mph at 1,000 feet.

B3479 was sent to No 19 Squadron on 9 June and, between tests, was flown operationally by LT G. S. Buck and CAPT F. Sowrey. Between them they brought down five enemy aircraft with the Spad 13. B3479 was later sent to No 23 Squadron, where it had a long career before being destroyed on 23 March 1918, after logging eighty-five hours nineteen minutes of flying time.

The test results with B3479 and its operational successes led the British to alter the Spad 7 contract with Kellner. The original order for 120 Spad 7s was changed, with half the aircraft now being Spad 13s. Additionally, a follow on order for an additional 100 Spad 13s was added to the contract. Deliveries were slow, because of the same problems being faced by the French. Additionally the French insisted on receiving half of Kellner's output, despite the fact that the aircraft were on order for the RFC.

The RFC had hoped that the first Spad 13s would be in service by August of 1917, however, the first aircraft did not arrive until November. Production continued to be slow and by March of 1918, the RFC had taken delivery of only sixty-one Spad 13s. In the event, a subsequent order for seventy additional Spad 13s was reduced to twenty-five.

No 23 Squadron, RFC was the sole British squadron to be equipped with the Spad 13. British pilots found the visibility from the cockpit to be restricted and reported that it was difficult to fly in formation. Because of these problems, the Spads were withdrawn and replaced with Sopwith Dolphins on 4 May 1918.

In French service, the Spad 13 proved itself to be a capable combat fighter. One of the earliest Spad 13s was flown by Georges Guynemer and it was in this fighter that he was lost on 11 September 1917. Other French aces that flew the Spad 13 included Rene Fonck, Georges Madon, Albert Heurtaux, Charles Nugesser, and Maurice Boyau. By war's end, some eighty squadrons had been equipped with the Spad 13.

This early Spad 13 (S.1929) reveals the rounded wing tips that were common on early production Spad 13s. An unusual feature of this particular Spad 13 is that all the wing ribs have been taped. This aircraft also has a rear-view mirror on the upper wing center section. (NASM)

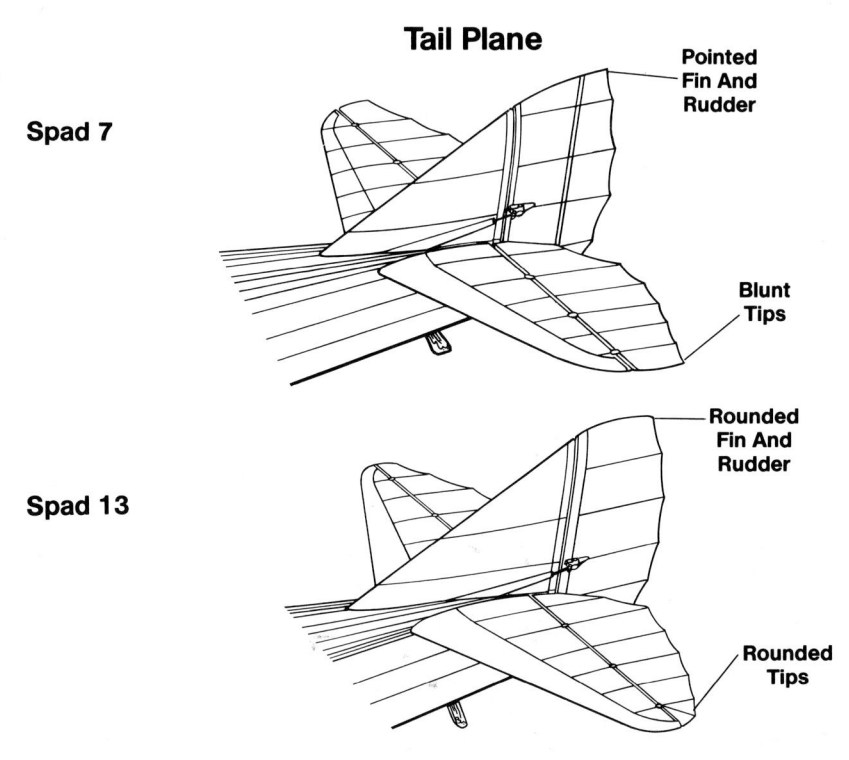

Tail Plane

Spad 7 — Pointed Fin And Rudder, Blunt Tips

Spad 13 — Rounded Fin And Rudder, Rounded Tips

American Service

The Spad 13 has been celebrated as the mount of such aces as CAPT Edward V. Rickenbacker (America's top scoring ace) and the dare-devil LT Frank Luke, Jr. The attention given to the Spad 13 often overlooks the fact that the United States Air Service (USAS) did not receive their Spads until the French had worked out all its problems. By the time the Spad 13 entered squadron service with the USAS, they considered it virtually obsolete.

Fifteen USAS squadrons were equipped with the Spad 13, augmented by the Spad 7s that remained in service. Many USAS Spads were re-equipped in the field with Marlin machine guns. In all, 893 Spad 13s were delivered to the USAS before the war ended. At one point, there was some interest expressed in building the Spad 13 in the United States, however, the war ended before these discussions were finalized and the type did not enter production. Unlike the earlier Spad 7, the Spad 13 was to be built only in France.

Spad 13s were also used by the 10e *Escadrille* of the *Aviation Militaire Belge*, however, apparently very few aircraft were delivered to the Belgians. A number of Spad 13s were also exported to Italy, and formed part of the 77th and 91st *Squadriglie*.

The Spad 13 continued in service with the *Aviation Militaire* until 1923. It also found post-war use in the air services of the United States, Poland, Spain, Czechoslovakia, and Japan — where it was known as the Hei 1.

The pilot of this early Spad 13 assigned to Spa. 3 waits while a ground crewman checks the guns before starting the engine. The pilot at the extreme left is believed to be Guynemer. (SHAA)

This Bleriot-built Spad 13 (S.2179) is believed to have been assigned to *Escadrille* N 561 based at Lido Aerodrome, near Venice, Italy. Pilot in the cockpit is thought to be Xavier Garros. (SHAA)

Wing Tips

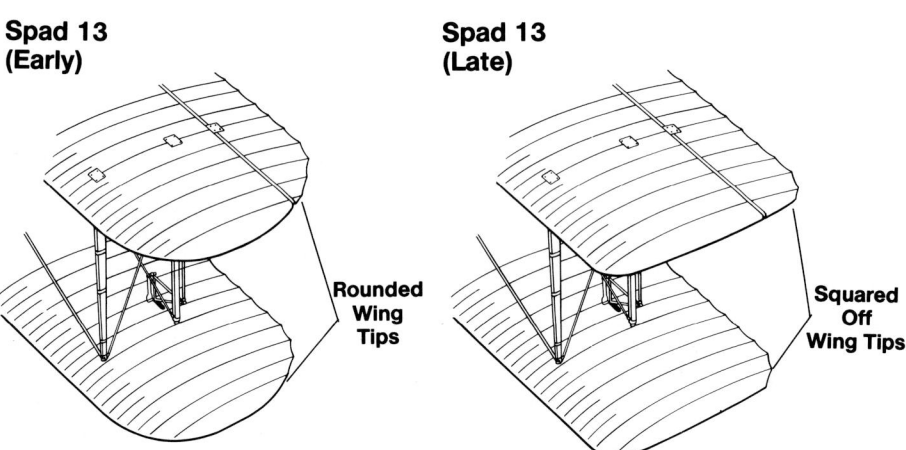

Spad 13 (Early) — Rounded Wing Tips

Spad 13 (Late) — Squared Off Wing Tips

These Spad 7s and 13s of an unidentified French *Escadrille de Chasse* parked on a forward French airfield are in the process of having the unit insignia painted on the fuselage. In the foreground is a Spad 7 (S.5190), while behind it is a Kellner-built Spad 13 (S.4567). (Musee de l'Air)

Guynemer, in flying coveralls, discusses his new Spad 13 with his father and other dignitaries at an unidentified French airfield. (Musee de l'Air)

Among the first French aces to fly the Spad 13 in combat was Guynemer, who was assigned Spad 13 (S.504) in September of 1917. Guynemer was lost flying this aircraft on 11 September. S.504 bears the Stork insignia of Spa. 3 (White with Back wings) and Guynemer's personal number, 2 on the fuselage side, however, it did not carry the name *Vieux Charles*. (Musee de l'Air)

This Spad 13 (S.1893), flown by ADJ Jacques Roques of Spa. 48, carries the later form of the unit's Fighting Cock insignia. The radiator cowling and the arc around the insignia are Blue, while Roques' personal number, 7, is in Red. (SHAA)

41

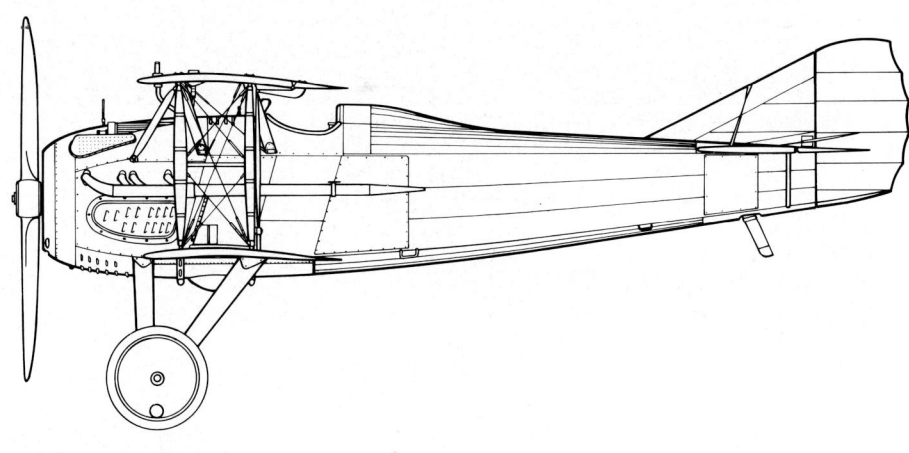

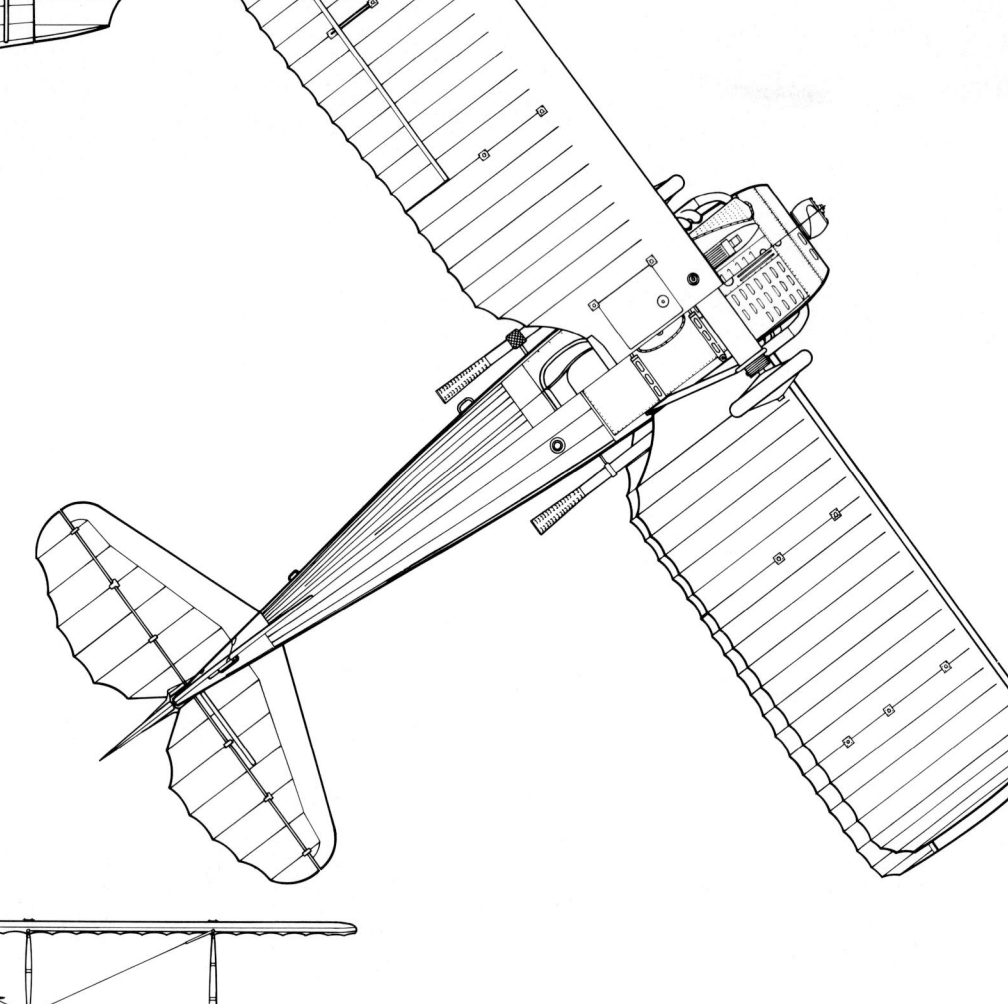

Specifications

Spad 13.C.1

Wingspan	26 feet 6 inches
Length	24 feet 3⅓ inches
Height	13 feet 1½ inches
Empty Weight	1,326 pounds
Maximum Weight	1,888 pounds
Powerplant	One 220 hp Hispano-Suiza 8BEa engine
Armament	Two 303 Vickers machine guns
Performance	
Maximum Speed	131 mph
Service ceiling	22,300 feet
Range	200-250 miles
Crew	One

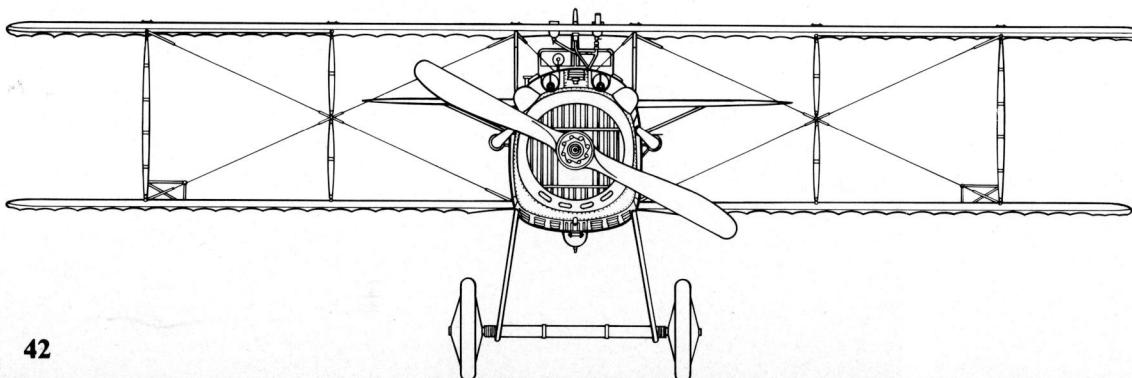

Jean Lucas alongside a late production Spad 13 of Spa. 97. The wooden fairing around the drag wire which runs from the top of the forward cabane strut to the top of the fuselage has the appearance of an additional strut. The individual number, 12, in White on the fuselage side was also repeated on the upper starboard wing. (SHAA)

This late production Spad 13, believed to have been assigned to Spa. 89, was brought down intact by LT Albert Hausmann of *Jasta* 13 in August of 1918. The aircraft was repainted in German markings and was test flown by *Jasta* 13. (Musee de l'Air)

Spads of Spa. 68 are neatly lined up on a French airfield. The two aircraft in the foreground are Bleriot-built Spad 13s while the aircraft in the background is a Spad 7. The French horn on the fuselage side is the unit insignia for Spa. 68. The horn and the individual aircraft number are repeated on the upper wing surface. (Musee de l'Air)

Charles Nungesser climbs out of his S.A.F.C.A.-built Spad 13. Nungesser's macabre personal insignia was carried on all his aircraft. His Spad also had tri-color diagonal stripes across the upper wing surfaces, and an off-center White triangle on the fuselage upper decking. This is also believed to be the sole combat aircraft flown by Nungesser that had a name, *La Verdier* (Greenfinch). (Musee de l'Air)

S.P.A.D., along with eight different subcontractors were involved in the production of the Spad 13. The aircraft factories of that period appear very primative when compared with a modern aero-space plant. It is difficult to believe that thousands of Spads were built under such conditions. (Musee de l'Air)

Components for at least two Spad 13s, including S.7542, prepare to leave the Kellner production plant. The aircraft were shipped by truck to a reassembly RGA where they were delivered to French fighter units. (Musee de l'Air)

ADJ Marius Blanc in the cockpit of his Spad 13. Blanc was assigned to Spa. 81, Fighting Greyhounds. The Greyhound insignia was one of the most attractive unit insignias used by French forces during the war. (Musee de l'Air)

An Italian pilot named Bertini of the 91e *Squadriglia* alongside his Spad 13 at Trieste, Italy, on 28 August 1918. The rampant griffin unit marking is believed to have been in Black, although it may also have been in Red. (Roberto Gentilli)

Italian ace Aldo Bocchese, by his Spad 13 of the 91e *Squadriglie*. Although a number of Italian Spads are believed to have been painted overall Dark Olive Green, this Spad retained its standard five-color French camouflage. (Roberto Gentilli)

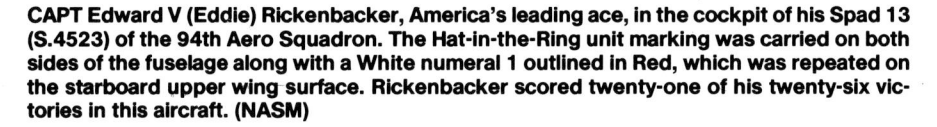

This Kellner-built Spad 13 (S.16541) was flown by MAJ Robert L. Walsh of the 22nd Aero Squadron, USAS. The fuselage number was Red, outlined in White, and the radiator cowling was Blue. This aircraft was a companion to the one which is now in the National Air and Space Museum in Washington. (Robert Sheldon)

CAPT Edward V (Eddie) Rickenbacker, America's leading ace, in the cockpit of his Spad 13 (S.4523) of the 94th Aero Squadron. The Hat-in-the-Ring unit marking was carried on both sides of the fuselage along with a White numeral 1 outlined in Red, which was repeated on the starboard upper wing surface. Rickenbacker scored twenty-one of his twenty-six victories in this aircraft. (NASM)

CAPT Robert Soubiran standing next to his Spad 13 (S.7714) of the 103rd Aero Squadron (formerly the *Lafayette Escadrille*). The Indian head unit marking was framed by a White-outlined Red diamond. The diamond was repeated on the horizontal tailplane, but without the White outline, and a tri-color stripe was added to the fuselage and radiator cowling. (NASM)

45

A group of mechanics and their dog alongside a Spad 13 of the 103rd Aero Squadron during 1918. This aircraft is believed to have been flown by LT Larne. (NASM)

This Black and White checked Spad 13 was flown by LT Robert W. Donaldson. The pilots of the 94th Aero Squadron quickly found out that the performance of their Spad 13s suffered because of the weight of the paint needed to apply these color schemes. (NASM)

While on occupation duty at Nieuwied during the Spring of 1919, several pilots of the 94th Aero Squadron, USAS, had their Spad 13s repainted in rather colorful paint schemes. This American Flag scheme was carried on the Spad 13 flown by CAPT Reed Chambers. The wings and fuselage were painted in Red and White stripes, while the nose and empennage were Blue with White stars. (Ernest R. McDowell)

Marechal des Logis Rene de Liniere alongside his Spad 13 of Spa. 103. Spa. 103 was part of the Allied Occupational Forces at Neustadt, Germany during the Spring of 1919. The Spad in the foreground is thought to be a Kellner-built aircraft fitted with a non-standard windscreen. (Musee de l'Air)

This Spad 13 was assigned to Spa. 84, which was also part of the French Occupational Forces stationed in Germany during 1919. (SHAA)

French officers prepare a Bernard-built Spad 13 for a demonstration flight in Japan. The Spad was part of the French Aeronautical Mission to Japan during 1919. The following year, the Japanese Army Air Force adopted the Spad 13 as its standard fighter under the designation, Hei 1. (SHAA)

This Spad 13 served with the 2nd Air Regiment of the Czechoslovakian Air Force during 1925. The national insignia under the wing is a representation of the Czech flag. (Zdenek Titz)

This Spad at Thionville during 1923, the last year the Spad served with French military aviation. By this time most of the Spads that remained in service were being used as advanced training aircraft. (Musee de l'Air)

Spad 17 and Spad 21

As more powerful versions of the Hispano-Suiza engine were developed and became available, Louis Bechereau made use of these engines to design several improved variants based on the Spad 13.

The first of these, designated the Spad 17.C.1 was designed around the 300 hp Hispano-Suiza 8Fb engine. The Spad 17.C.1 was based on the Spad 13 and its dimensions were identical. Structurally, however, it differed significantly with the earlier Spad 13. Its wings were strengthened, the fuselage was bulkier and was fully faired with numerous stringers. The larger horizontal stabilizers were braced by two struts mounted below the stabilizer, ahead of the elevator hinge line. Auxiliary flying wires were also attached from the undersurfaces of the lower wings to special attachment points on rearward extensions of the undercarriage V shaped struts.

The first Spad 17 began testing in April of 1918. The flight tests revealed a top speed of 217 kph (135 mph) at 2,000 meters (6,560 feet) while its speed at 5,000 meters (16,400 feet) was 201 kph (125 mph). Time to climb to 5,000 meters was seventeen minutes, twenty-one seconds.

The Spad 17's performance was marginally better than the Spad 13 and a limited production run of twenty aircraft was ordered, presumably as a pre-production test batch (as had been done with the Spad 13). A number of these aircraft became operational in June of 1918 being assigned to the famed *Les Cigognes* (Storks). Among the pilots who flew the Spad 17, late in the war, was France's leading ace, Rene Fonck.

The Spad 21 was a contemporary of the Spad 17. The aircraft was externally nearly identical to the Spad 17, except for having upper and lower wings of equal span and chord with ailerons mounted in both wings. The ailerons were controlled via a connecting link housed in a fairing behind the outer interplane strut on each wing. It was hoped that this arrangement of double ailerons would improve the Spad's maneuverability, however, tests conducted during October of 1918 revealed that its performance was actually poorer than that of the Spad 17. Although the Spad 21 was slightly faster in level flight, its rate of climb was slower and its maneuverability did not show the desired improvement. Based on these test results, the project was abandoned in November of 1918. The Spad 21 reportedly had a top speed of 221 kph (137 mph) at 2,000 meters and 205 kph (127 mph) at 5,000 meters. Time to climb was five minutes forty seconds and eighteen minutes eighteen seconds, respectively.

This is one of twenty Spad 17s (S.733) built during the spring of 1918. The Spad 17 had a faired fuselage, strengthed wing, and enlarged fin and rudder. The rearward extension of the undercarriage strut and the additional flying wire running from the landing gear to the underside of the wing is visible just behind the port wheel. (Musee de l'Air)

This is one of two Spad 21s (S.777) built and tested by the French. The aircraft had a bulkier fuselage than earlier Spads and carried ailerons (with inverse taper) on both the upper and lower wings. The metal fairing on the outboard rear wing strut enclosed the aileron acutating rod. (Musee de l'Air)

Spad 14 and 24

The Spad 14 was basically a seaplane version of the earlier cannon armed Spad 12. The Spad 14 was similarly armed, carrying a 37MMmm Hotchkiss cannon firing through the propeller hub and a synchronized Vickers machine gun on the upper fuselage in front of the cockpit. To compensate for the added weight of the floats, the wings of the Spad 14 were lengthened. Wing span was seven feet thirteen and a quarter inches longer than those of the Spad 12 (32 feet 1¾ vs 26 feet 3 inches) with the wings being rigged as a true two-bay wing. The area of the fin and rudder was also increased over the Spad 12 to improve lateral stability.

The prototype Spad 14 first flew on 15 November 1917. It had a top speed of 205 kph (127 mph), which is believed to be a record for a seaplane at that time. A total of forty Spad 14s were built by Pierre Levasseur and these aircraft saw limited service with the French *Aviation Maritime*. After the war, one Spad 14 was flown in a seaplane race at Monaco during 1920.

The Spad 24 was a landplane conversion of the Spad 14 and was intended to be used as shipboard fighter. It first flew on 5 November 1918, too late to see service and further development was abandoned.

Spad 22

The most unusual development of the Spad single-seaters was the Spad 22. It utilized the fuselage of the Spad 17 with radically redesigned wings. The upper wing had a pronounced sweep back, while the lower wing had an equally pronounced forward sweep. To accommodate this arrangement, the interplane bracing had to be drastically redesigned. The tail surfaces were also redesigned with horn-balanced elevators. The intent of this wing arrangement was to improve the pilot's visibility from the cockpit, one of the biggest complaints of earlier Spads.

The Spad 22 was being test flown at Buc during November of 1918, powered by a 300 hp Hispano-Suiza 8Fb engine driving a two blade propeller, when the war ended. It proved unsuccessful and, with the end of the war, further development was abandoned.

The Spad 22 featured a swept back upper wing and swept forward lower wing, which was intended to improve the pilots visibility from the cockpit. The aircraft also featured radically redesigned wing interplane bracing struts. (IWM)

A crewman of the racing Spad 14 prepares to attach the hoisting cable to the aircraft to lift it clear of the water. This aircraft was participating in the seaplane races held in Monaco during 1920. (Russell Hansen)

This Spad 14 was basically a seaplane variant of the earlier cannon armed Spad 12. A number of these aircraft served with the French *Aviation Maritime* during 1918. (Russell Hansen)

Eagles of The Great War, 1914 - 1918

1046 Albatros

1098 Fokker Dr 1

1123 BE 2

1137 Bristol Fighter

1158 Fokker Eindecker

1164 de Havilland D.H. 9

1166 Fokker D VII

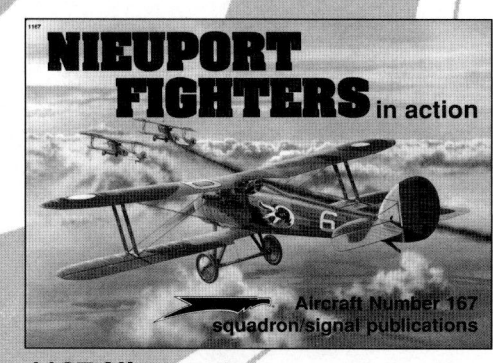

1167 Nieuports